BIOPOLITICS OF SECURITY

Taking its inspiration from Michel Foucault, this volume of essays integrates the analysis of security into the study of modern political and cultural theory.

Explaining how both politics and security are differently problematised by changing accounts of time, the work shows how, during the course of the seventeenth century, the problematisation of government and rule became newly enframed by a novel account of time and human finitude, which it calls 'factical finitude'. The correlate of finitude is the infinite, and the book explains how the problematisation of politics and security became that of securing the infinite government of finite things. It then explains how concrete political form was given to factical finitude by a combination of geopolitics and biopolitics. Modern sovereignty required the services of biopolitics from the very beginning. The essays explain how these politics of security arose at the same time, changed together, and have remained closely allied ever since. In particular, the book explains how biopolitics of security changed in response to the molecularisation and digitalisation of Life, and demonstrates how this has given rise to the dangers and contradictions of twenty-first century security politics.

This book will be of much interest to students of political and cultural theory, critical security studies and International Relations.

Michael Dillon is Professor of Politics and International Relations at Lancaster University. He has written extensively on international political theory, continental philosophy, security and war, and cultural research.

PRIO New Security Studies
Series Editor: J. Peter Burgess, PRIO, Oslo

The aim of this book series is to gather state-of-the-art theoretical reflexion and empirical research into a core set of volumes that respond vigorously and dynamically to the new challenges to security scholarship.

The Geopolitics of American Insecurity
Terror, power and foreign policy
Edited by François Debrix and Mark J. Lacy

Security, Risk and the Biometric State
Governing borders and bodies
Benjamin J. Muller

Security and Global Governmentality
Globalization, governance and the state
Edited by Miguel de Larrinaga and Marc G. Doucet

Critical Perspectives on Human Security
Rethinking emancipation and power in international relations
Edited by David Chandler and Nik Hynek

Securitization Theory
How security problems emerge and dissolve
Edited by Thierry Balzacq

Feminist Security Studies
A narrative approach
Annick T. R. Wibben

The Ethical Subject of Security
Geopolitical reason and the threat against Europe
J. Peter Burgess

Politics of Catastrophe
Genealogies of the unknown
Claudia Aradau and Rens van Munster

Security, the Environment and Emancipation
Contestation over environmental change
Matt McDonald

Securitization, Accountability and Risk Management
Transforming the public security domain
Edited by Karin Svedberg Helgesson and Ulrika Mörth

Commercialising Security
Political consequences for European military operations
Edited by Anna Leander

Transnational Companies and Security Governance
Hybrid practices in a postcolonial world
Jana Hönke

Citizenship and Security
The constitution of political being
Edited by Xavier Guillaume and Jef Huysmans

Security, Emancipation and the Politics of Health
A new theoretical perspective
João Nunes

Security, Technology and Global Politics
Thinking with Virilio
Mark Lacy

Critical Security and Chinese Politics
The Anti-Falungong Campaign
Juha A. Vuori

Governing Borders and Security
The politics of connectivity and dispersal
Edited by Catarina Kinnvall and Ted Svensson

Contesting Security
Strategies and logics
Edited by Thierry Balzacq

Conflict Resolution and Ontological Security
Peace anxieties
Edited by Bahar Rumelili

Biopolitics of Security
A political analytic of finitude
Michael Dillon

No one writing about security issues is as philosophically astute, conceptually innovative, politically attuned, and eloquent as Michael Dillon. Everyone in academic, journalism, and policy communities concerned with security thinking should read *Biopolitics of Security*.

Michael J. Shapiro, University of Hawai'i, Manoa

BIOPOLITICS OF SECURITY

A political analytic of finitude

Michael Dillon

Routledge
Taylor & Francis Group
LONDON AND NEW YORK

First published 2015
by Routledge
2 Park Square, Milton Park, Abingdon, Oxon OX14 4RN

and by Routledge
711 Third Avenue, New York, NY 10017

Routledge is an imprint of the Taylor & Francis Group, an informa business

British Library Cataloguing in Publication data
A catalogue record for this book is available from the British Library

Library of Congress Cataloguing in Publication data
A catalog record for this title has been requested

ISBN: 978-0-415-48432-9 (hbk)
ISBN: 978-0-415-48433-6 (pbk)
ISBN: 978-1-315-72501-7 (ebk)

Typeset in Bembo
by Out of House Publishing

MIX
Paper from responsible sources
FSC® C013056

Printed and bound in Great Britain by
TJ International Ltd, Padstow, Cornwall

For Laurence Montagu

By the time that we have discovered that things are at a given juncture, they have already changed several times. Hence we always perceive events too late, and politics must always foresee, as it were, the present.

(Turgot)

For all their boasting, practical men do not know either men or the world; they do not even know the reality of their own works. [If they could return to life], the geniuses of pure politics, the *fatalia monstra* recorded in histories, would be astounded to learn what they have done without being aware of it, and they would read their own past deeds as in a hieroglyph to which they had been offered the keys.

(Benedetto Croce)

CONTENTS

ACKNOWLEDGEMENTS

I would, first, like to thank the editors for their extraordinary forbearance and support.

Peter Burgess (PRIO), my series editor, commissioned the project and never said a cross word about how long it has taken me. He could have been forgiven for saying many.

Alexander Hope edited these essays as he did those that made up my previous book, *Deconstructing International Politics* (Routledge, 2013). Unaccountably he was missing from the acknowledgements to that book and so I doubly thank him now for his help. He began as an editor and ended a valued intellectual friend and colleague.

My thanks also to the following publishers for permission to republish material from the following essays:

With Luis Lobo-Guerrero, 'Biopolitics of Security in the 21st Century', *Review of International Studies*, 2008 (Cambridge University Press)
'Underwriting Security', *Security Dialogue*, 2008 (Sage)
'Governing Terror', *International Political Sociology*, 2007 (Wiley)
'Security, Race and War', in Michael Dillon and Andrew Neal, eds., *Foucault on Politics, Security and War*, 2008 (Palgrave Macmillan)
'Politics of Truth and Pious Economies', 2013 in Sandro Mezzadra, Julian Reid and Ranabir Samaddar, eds., *The Biopolitics of Development: Reading Michel Foucault in the Postcolonial Present* (Springer Press).

One of the great joys of academic life is the friendship to be made along the way. I have been very fortunate in this respect. I would especially like to thank Ward Blanton, Antonio Cerella, Andy Dobson, Andrew Gibson, Charlie Geere, Paolo Palladino, Andrea Rossi, Yvonne Sherwood, and Samuel Weber. I wish to give special thanks to Luis Lobo-Guerrero (Groningen) who helped me to begin, and has followed the project throughout. I thank him also for permission to use an essay we wrote together in the early days of its formation, and not least for his continuing faith in me. Julian Reid (Rovaniemi) is a singular friend. He has the special gift of inspiring the unruliness in me that helps me reach out to other thought. Bulent Diken (Lancaster and Mardin) talked through the entire book with me and together with Arthur Bradley (Lancaster) has kept me alive here in Lancaster. My thanks and best wishes also go to Edogan Sima, Alev, Drogan and Zeynap for taking care of me in Istanbul. Finally, I would like to thank my old mentor from Dalhousie University Canada, Professor Dennis Stairs. Now that I have also become an old teacher, I understand what a surprise it is to be thanked for the good one might have done without even knowing it. Dennis was a wonderful teacher to me. I would not have become the Professor I became if he had not first encouraged me to start out.

Since these acknowledgements are a celebration of friendship, I wish to dedicate the book to my oldest friend, Laurence Montagu. Laurence and I grew up together in working-class Liverpool. His family was my family. Laurence went on to be a nationally decorated and acclaimed high school Headmaster. He devoted his life to improving the prospects of those without privilege, and championing their welfare. He was justly celebrated for it throughout English secondary education. It was Laurence who taught me that you could not teach anyone anything unless you first treated her or him with love and respect. For that reason, I know, thousands filled Gloucester Cathedral at his passing.

INTRODUCTION

> This great Leviathan which is called the state, is a work of art; it is an artificial man made for the protection and the salvation of the natural man.
>
> (Hobbes 1960: lxiv)

> The Baroque refers not to an essence but rather to an operative function, to a trait.
>
> (Deleuze 1993: 3)

I thought I had finished this book long ago. It was to be a series of previously published essays exploring different expression of what Michel Foucault called biopolitics. I was not so much interested in considering how Foucault's account of biopolitics differed from other accounts of biopolitics, most notably those of Agamben, Esposito, and Negri and Hardt. It was not a matter of dismissing these accounts. I was simply not that interested in confining myself to exegesis, or to comparative analysis. Inspired by Foucault, I was more interested, like them, in continuing to put his insights to work. I was especially interested in updating Foucault's account of biopolitics to encompass how the digitalisation and molecularisation of Life had changed the biopolitical mission to 'make live'. And I was especially interested, also, in developing his account of the biopolitics of security. Or, rather, as Foucault put it, of biopolitics as a collection of changing *dispositifs de sécurité*. So far, so good.

Then things changed. It was not so much that I was dissatisfied with the essays. They were, of course, limited in the many ways in which all essays are limited, and no doubt, also, in many ways specific to these essays as well. I was bound to be dissatisfied. That wasn't the problem. Having been invested in Foucault's lecture courses for some considerable time, I had made the 'mistake' of re-reading Foucault's magnum opus *The Order of Things*. It is a brilliant and somewhat neglected book. Reading it once more put my own book back years; and precisely because *The Order of Things* is so clever and prescient. In the penultimate chapter in which Foucault analyses the aporias characterising the modern episteme – of how Kant had given epistemic form to finitude, anchoring it in the figure of Man – I was especially captured by that single expression, 'giving concrete form to finitude'. What, I thought, is meant by finitude here? Surely finitude is just the fact that we are born and that we die. But, of course, not. There are many different accounts of finitude. Why, also, must concrete form be given to finitude? And how? What could this mean?

I consequently paid renewed attention to Foucault, in general, and *The Order of Things* in particular. What was the finitude that Foucault was talking about? How did it matter politically? What relevance did finitude have to politics of security, in general, and to the biopolitics of security in particular? Better put, how were both politics and security problematised when they were addressed through the optic of finitude, whatever this finitude was taken to be? What form, or forms, does concrete form take? Is there such a thing as concrete political as opposed, for example, to concrete epistemic form, which is what I took Foucault to mean when discussing Kantian epistemology? And why, especially, was Foucault so fascinated with the baroque? I had never given any serious thought to the baroque. I assumed that it referred to a period and to an aesthetic, and had not linked that closely to the question of temporality in the form of finitude, or with politics, government and rule as these came to be problematised through the temporal enframing of time as finitude. How, then, were time, this finitude and politics in general, as well as politics of security in particular, linked?

I could have done without this. But then I discovered that I couldn't write the book without posing and answering these, and other related, questions, at least to my own satisfaction. I had already worked out how biopolitics of security differed from geopolitics of security. One revolved around the properties of Life understood in terms of species existence, while the other revolved around the properties of the sovereign territorial state. That story is part of the story that gets told throughout the essays that follow. That was, however, the easy bit. What had 'finitude' got to do with it? And what finitude were we, was Foucault, talking about?

Working this out is what took the time. It also took me a long way from Foucault and, then, it seemed inevitably, back to Foucault. The journey was exhilarating, but it long delayed the book, and radically changed the enframing of it. I came to see the essays in a new light, and the whole enterprise began to concern the modern problematisation of politics, government and rule as such, rather than the biopolitics of security as one expression of it alone. I also came to accept, at face value so to speak, Hobbes' account of how the seat of power is centred in artifice, to connect that to the baroque of his time, as well as ours, and then to appreciate, more generally, how baroque artifice characterised the very character of modern politics as such, the politics of security especially. For it became clear throughout this process of reading and reflection that the baroque was not a mere period, nor was it confined to the aesthetic, to art and literature. It was a shifting field of emergence, formation and problematisation in which the problematic of politics, government and rule was cast in a novel and dynamic way, preoccupied with the manipulation of appearance. Moreover, it was a field of formation and problematisation comprised of many changing points of application, of many characteristic and inventive modes of expression, as well as surfaces of friction, that were themselves a response to something else.

If the baroque engendered a certain political anthropology, for example that classically of the sovereign tyrant and martyr, prey also to the court plotter, depicted in the German tragic dramas (*Trauerspiel*) of the late seventeenth century, analysed by Walter Benjamin, it was not only a field of formation and problematisation, it was one characterised, also, by the absence that the retreat of the Christian God into the otherworldly obscurity of his inscrutable majesty had left in accounts of nature, human nature, and time itself. This *deus absconditus* (Blumenberg 1985) resulted in a profound, historically engendered lack, the absence of any transcendent intelligibility ordering the universe in general, and human affairs in particular. The lack had to be filled by human artifice. Filled, not only in the field of knowledge, the modern episteme, or even also in the field of social and political organisation. The point applied to the very constitution of modern individual self-hood.

There was, in short, no nature, no human nature, that was not comprised historically, here, of this lack in the nature of nature as such. Culture, the whole gamut of human artifice, was summoned to bridge the gap. But at the same time it was of course fated, fated to fail since there is no bridging the gap. The artifices of human design and desire play it out, but they cannot resolve it. In other words, the baroque was itself a field of emergence, formation and problematisation responding to a lack in the time bequeathed to it by the withdrawal of the Christian God. It was this historically emergent temporal lack – of the advent of an understanding of time, or history, bereft of extra human guarantees as to its intelligibility or teleology – that had been precipitated by the loss of faith, politically and socially at least, in the promised resolutions offered by Christianity's salvation history.

Here was the link to the finitude to which Foucault referred in *The Order of Things*.[1] Here, also, was the link to the business of having to give concrete form to a finitude which was no longer understood to be informed and underwritten by a divinely ordered, salvational, order of things. Since there was no guaranteed form in nature, theistically or naturally, other than that which positivistic knowing was to draw from it through the astonishing power of the epistemic, scientific and technological devices that demand it show itself in forms consonant with the ways in which it was summoned to show itself, concrete form was brought to it, or brought out of it, in terms set by the mechanisms through which it was addressed.

With this additional point. The baroque is no fixed field of formation and problematisation. It is a changing manifold of formation and problematisation responding to an historically induced sense of a profound lack in nature, specifically, as it were, in human nature, that was bequeathed by the dissolution of a soteriological understanding of finitudinal time. This required that the lack, thereby opened up, be bridged by the genius of human artifice. Not by human artifice as such, as if it transcended the ways in which it is manifested historically. The baroque was distinguished less by a reliance upon artifice as such, therefore, than a reliance upon the radically changing forms of artifice introduced, from the sixteenth and seventeenth century onwards, by successive scientific and epistemic revolutions, and their allied rules of truth and truths of rule, especially those concerning politics, security and government (Gal and Chen-Morris 2013a; and Gal and Chen-Morris 2013b). The pathos of modern mourning for this lack, the absence of something towards which it expresses radical incredulity, has an odd ring to it, since what seems most characteristic of the baroque character of modern times is not lack but the very pluripotency of its morphological capacities, its apparently infinite ability to experiment creatively with form and renew finite appearance economically and politically as well as scientifically.

What inspired the political anthropology of the baroque was, therefore, the political temporality to which it gave expression in the field especially of politics, government and rule, as well as of representation in the arts, science and literature (Gal and Chen-Morris 2013a; and Gal and Chen-Morris 2013b). Indeed the entire ordering of representation was, thus, set in play in ways not only different from but also alien to Christian rules of truth and truths of rule, because the representation of representation, as such, had been problematised in novel ways. This was the lesson that Foucault taught via his reading of Velasquez's painting *Las Meninas* with which he introduced *The Order of Things.*

Politics, government and rule are always also correlated with, align themselves to, the properties of whatever is to be governed and ruled. Since the defining property of that which was now to be ruled was, henceforth, defined by the lack supposed at its centre, by a lack once filled by the soteriological account of human finitude offered by Christianity, the baroque is that field of formation and problematisation, and of application, which explicitly negotiates and exploits this gap by mobilising human artifice within the theatre of opportunity of necessity and contingency that it presents; of the very necessity of contingency, and contingency of changing necessities. This chiasmus of terms, less their binary opposition than their endless intersection and reversibility, is itself, also, a characteristically baroque response to a world rendered more and differently contingent, and therefore in its own manifold ways also, more necessary, than hitherto imagined. The baroque sensibility does not so much expect to heal this rift in time and nature, as conceived from this historical perspective of the loss of the Christian God, as construe it as a stage, a *theatrum mundi*: 'There was ultimately no more visible or influential way to take part in baroque social principles than in theatrical representations' (Maravall 1986: 234). Here, then, was an opportunity to put human artifice on display and to work as the very means by which natures of every description, having no inner or external guarantee, are nonetheless capable of having the marvels, mysteries and predicaments that they pose played out through human enframing, intervention and design. Here was a new real. Its realism was 'never a question ... of pure realism. The remains of idealism, of Neoplatonism, appeared everywhere.' (Maravall 1986: 258).

One final important point. This lack, this so-called deficiency, was a consequence of the withdrawal, absence, or, otherwise radical incredulity that began to be expressed towards narratives, liturgies and belief systems once premised on the proposition that the order of things revolved around a very particular, historically variable, expression of divine order. It is the gap left by the dissolution of the social and political priority accorded to precisely that historical field of formation in religious, social and political affairs alike, that leaves the gap. It may concern the ontological – specifically onto-theology – but it was, in fact, historical. The peculiarity of the modern is that, so self-consciously aware of itself as historical, it seems to mourn the loss of this extra human guarantor as if it were ontological, rather than an historical, phenomenon.

As this new historical play of rules of truth and truths of rule gathered power and pace, so also, however, were the 'natures' in play transformed. This theatre of 'deficient' natures is thus unstable precisely because, acting into it, human artifice transforms the very taking place of natures themselves, their very appearance. And does so in radically contingent, unexpected and ultimately, it seems, vertiginously unmanageable ways.[2] Its military and political spectacles, especially, cast the human self in terms of spectator, artificer and, ultimately also, mere disposable stuff. In political terms this very staging, the artifice to which Hobbes refers, for example, is designed to render human beings more governable via their very spectatorship on the one hand, and their artfulness on the other. Consider how, in the frontispiece to *Leviathan*, for example, the spectators' gaze is transfixed by the visage of the sovereign, and that the sovereign is a visage for which they all fall. They seem stupefied by it. And they are meant to be so. Transfixed by the abstract gaze of the sovereign – its eyes are empty and unseeing, its visage a death mask – they fall for its promise of security. Put differently, the very organisation of the theatre of politics around the gaze of the sovereign is designed to keep subjects fixated in this way. For that way they are rendered governable; at least governable by these means, because subjectified by its ordering of things.

Only marginally does Hobbes accept that the relationship is reversible. But the acceptance is nonetheless there, as an integral part of the rule of truth and truth of rule of which it is comprised. Leviathan is a veridical apparatus and Hobbes a certain new form of political truth-teller. There is, however, no reason in principle why, in the Hobbesian order of things, spectators cannot themselves become artificers. If they feel their security threatened they are in principle, according to Hobbes, allowed to become so. No reason why those domesticated through the play of power may not, therefore, also turn that play against itself in newly inventive ways.

Here, then, a further peculiar feature of the baroque becomes evident. It is not simply a formation of truth and rule. It is a problematisation of truth and rule to which many different expressions may be given. One, moreover, in which the very staging and playing-out of truth and rule is evidently not only changeable, but also reversible. For there is no natural law, or nature of natures, that specifies that spectators cannot become artificers, or that artificers may not in turn become spectators; in the ways, for example, that contingency and necessity might be reversed, or in the way that, to give a different and more revealing example to which Foucault gives the name politics of truth, every rule of truth expresses itself as a truth of rule, and every truth of rule expresses itself as a rule of truth (there is, of course, always more than one rule for every truth, and more than one truth for every rule). Falling for the sovereign staging of power plays in this way, submitting to the shock and awe of the sovereign's gaze, translates power into authority, sovereignty into legitimacy. But it may not only do so. It may also witness the translation of authority back into its *an–archē*, legitimacy back into its in-credulity, as, for whatever reasons, the sovereign drama loses its capacity to persuade, enchant or stupefy; there is, of course, always an enchantment to persuasion, as there is persuasion in enchantment, since in-credulity comprises each. It is, in short, always possible for authority and legitimacy to lose their credence. There may be no external guarantees in which human judgement can be grounded. But who says spectators do not judge, do not refine their capacities to judge as they spectate? Even as spectators, human beings develop the capacity to exercise and refine their capacities to judge.[3]

Here, then, if dimly, I began to understand the finitude to which Foucault referred, how the need to give concrete political as well as epistemic form to this post-Christian finitude had arisen, and some of the characteristic ways in which the processes, thus set in play, proceeded to play themselves out. The baroque as field of emergence, formation and problematisation was, therefore, itself both a response to, and a certain engendering expression of, the temporality – factical finitude – that gave rise to it. If there was, historically speaking, a lack of naturally given form in this finitude, then of course, concrete form had to be given to it. In *The Order of Things*, Foucault explored how knowledge came to give concrete epistemic form to finitude in the form of Kantian epistemology, itself reliant upon the strategic figure of Man. Specifically also, however, if no natural political order obtains in this finitude, then concrete political form has to be given to it as well. Moreover, to the degree that this account of finitude is one in which we moderns continue to exist, and our politics, government and rule continue to be problematised, then our politics remains predominantly baroque.

I enter the caveat here that it would not remain baroque if time were construed in terms other than of the infinity of finite things currently obtaining in modern factical finitude. Herein, I think, lies contemporary interest in messianic time. Messianic time is construed as the time that remains. It is a different kind of 'now' time, one whose taking place takes place in relation to an ethical and political order of things inspired by what Derrida, for example, calls a politics to come (Bradley and Fletcher 2013; Dillon 2012). Such a time is not premised on the arrival of a Messiah. It posits the possibility of an order of things whose principle of formation is the 'as if' of anticipating its arrival at any time. But such a time is not the time at issue here, and I was yet to specify what specified this finitude as the finitude that it was.

And then it dawned. I already knew this, so I don't know why it took so long to recall it. Finitude has a history. It is figured in different ways. Christianity, for example, has an account of finitude. I call it soteriological finitude because it construes the nature of human finitude, in particular, from the perspective of the

Christian history of salvation. All things in nature from this perspective are understood to exist *sub specie aeternitatis.* This does not mean that they last forever. They don't. They are, of course, finite. They begin and they end. However, God's creation is ultimately prophesised to be impermanent. All things will end. And, when they do, all things will be gathered back into God's eternal presence; except, of course, those that fail the enigmatic standards of salvation. Their eternity is prophesised to be different. It is in this sense, then, that human beings especially are regarded as *sub specie aeternitatis.* That is to say their very finite appearance is a function of the eternal. Ultimately, their government revolves around it as well.

Not so with the finitude to which Foucault was referring. Not so with modern finitude, to which I therefore give a different name, that of factical finitude. Here, things do not exist *sub species aeternitatis.* They exist *ad infinitum.* Time is not soteriological, history is not salvation history. There is, simply (factically), an infinity of finite things. Infinity is not eternity. If eternity is the cause and ground of things existing *sub specie aeternitatis*, infinity is the correlate of the finite. Politics, government and rule are problematised theologico-politically from the perspective of soteriological finitude. Not so with factical finitude. Politics, government and rule are problematised in terms of the infinite government of finite things. The infinity of finite things knows no eschatology in the traditional sense of an ending to time combined with the advent of a different, a better, time to come.

Factical finitude is, of course, saturated with endings and a profoundly imminent as well as immanent sense of an ending. But it is not that ending. There is the artifice that creates and manipulates appearance. But there is no redemption from the necessity of engaging in the contingent exercise of appearance making. In as much as the formulation of the modern task of politics, government and rule is, therefore, construed foundationally as keeping formless, lawless, chaos at bay, the task of politics, government and rule is foundationally considered to be that of katechontic restraint. Hence, foundationally speaking, modern politics is a politics of security before it is a politics of anything else. Thus, if you desire a different politics, you have to posit a different order of time.

Additionally, I had intuited that these, and other characteristic features of the baroque imperatives continually played out in modern political institutions and practices, the changing manifold of power relations and processes of subjectification of which modern politics is comprised, often times explicitly theatrical, were most intensely and most extensively played out in its politics of security. And why? Because the retreat of the Christian God into the mystery of its all-powerful majesty deprived historical human being of the government that it offered in the form of the security of salvation. Long concerned with the correlation of the geopolitics of security with the biopolitics of security, I also began to consider seriously how they were related analytically, as well as historically, in as much as they were clearly among the most important mechanisms by which concrete political form came to be given to the factical finitude to which Foucault was referring: the finitude to which the baroque was a field of emergence, formation and problematisation, as well as experimentation in representation and form; the political temporality within which politics, government and rule came to be problematised throughout the times we call modern, and the changing processes that we call modernisation. If geopolitics was, however, governed by the strategic figure of sovereign reasoning Man, that of biopolitics was governed by the strategic figure of species existence, Life. Different strategic figures signalled different expressions of our modern politics of security.

The emergence of my problematic did not take place in an orderly research fashion, much less automatically as the application of some supposed Foucauldean template. It seemed, instead, to proceed through long periods of reading that resembled a crab-like walk through foreign textual landscapes punctuated by flashes of insight as to where I was, and in what direction I was heading. Through a process of iteration I formulated my questions better and saw connections, again both historical and analytical, that I had not seen before. It seemed like this might go on forever – it could have done, and some of my friends feared that it might – but I nonetheless thought I was getting somewhere. A sense of an ending began to form. And now I have to end.

What follows is a record of an enquiry I had not anticipated, into some things I did not know or understand before, and this is the result. I wish that it were better, more organised and systematic than it is. But, then, I already knew enough to know that the best is the enemy of the good, so I began to hope that this would be good enough. Whether or not it is, the reader will decide. One of the joys of writing is this. You have no idea who might read what you write, or what they will make of it. You simply cannot know. One of the other joys of writing is the discovery that the writing re-does you, as you do the writing. This I do know. So. I take my departure from my text a changed intellect. I like that. It has refreshed me and has inspired new interests. I am thankful to it.

Such enterprises are, however, also very selfish engagements. As you are overtaken by your obsession to the neglect of other things, and people, you can only beg forgiveness. I also know that I should have had a specific audience in mind, or some side I should side with, and set out to satisfy their concerns so that the book could be marketed more effectively to them, and be bought by librarians able to assign it to its proper space in the library. I always find myself in the wrong place in the library, however, reading material that seems to me to belong elsewhere. There is no established index to that elsewhere, because the elsewhere was whatever problematic had brought me to this or that unfamiliar place in the stacks in the first place. No Dewey system for codifying the intellectual interests of the wayward, the obsessive and the curious. Foucault's warrant for his work was his curiosity. He taught me to see and embrace the fact that it is also mine.

In what Foucault called the government of men and things, the order of things never adds up, or may always be added up differently. My curiosity lies there, in the processes by which the government of men, women and things are obsessively audited and assayed in order to make them add up, this way or that. And in the posturing and perfomativity, in the arbitrary as well as systemic violence and injustice, both petty, monumental and vile, that are committed in the name of securing the infinite government of finite men, women and things. I am curious about all of this, but there is of course much more left out than included in the account that follows. I nonetheless hope that what is included will be some compensation for what is omitted. Here, then, is a summary account of what is included. It focuses on the book's analytic more than the substance of the chapters. It does so because it is the analytic that matters in the end. It attempts some response to all the questions that Foucault's 'giving concrete form to finitude' prompted. It is, so to speak, the moral of the story that the chapters seek to tell

It helps to stage this opening by recalling some key points from Walter Benjamin's classic study of German Baroque drama of the late seventeenth century (*The Origin of German Tragic Drama* [*Trauerspiel*], 2009 [1925]). Benjamin acutely describes 'the typology and political anthropology of the [early] baroque' to which the plays gave expression (Benjamin quoted in Weber 1992: 9). My book moves from the political anthropology that Benjamin describes to the political temporality of factical finitude which is the condition of possibility to which the political anthropology of the *Trauerspiel* gives expression. As he encounters the question of early modern sovereignty addressed in the plays, it becomes clear that, for Benjamin, the question of sovereignty is not simply a theme of the German baroque. Like the baroque itself, the sovereignty dramatised there is what Foucault would have called a theoretical and methodological, as well as practical, space of problematisation. Benjamin's typology of the political anthropology that furnishes the actors, whose predicaments are dramatised by the plays, centres on three main figures: the tyrant, the martyr and the plotter. Suspended between the infinity of sovereign responsibility, and the finitude of his capacity to govern (decide) now posed by the shift from soteriological to factical finitude, the baroque sovereign finds himself facing a sovereign decision that is both necessary and impossible. He responds, first, by becoming a tyrant. The sovereign seeks the infinity of power that will enable him to overcome the finitude of his judgement: 'and yet the more he has, the more he demonstrates his incapacity to arrive at an effective decision' (Weber 1992: 15).

His failure is inevitable, not because he lacks this or that capacity, this or that information, but because sovereignty, itself, is fatally flawed by the very temporality that gives rise to it; that of the factical finitude bequeathing it the impossible task of exercising infinite government over the infinity of finite things. It is an aporia centrally comprising a very specific, definitive and irresolvable problem. To be sovereign, to become sovereign, requires the exercise of a strategic calculus of necessary killing. This is no simple matter of who shall die, politically, but how much killing is enough to institute, restore or otherwise secure the institution of the stable juridico-political order for which the sovereign is responsible. No strategic calculus of necessary killing is humanly available. Asked to determine how much killing is enough, sovereigns can only ever answer more or abandon the attempt be sovereign. Immobilised by this expression of the aporetic decision that defines him, the sovereign is suspended in the unbridgeable gulf between the infinite responsibility to decide and the finite capacity to do so. In consequence he is transformed into a martyr. The martyr figure is typically described in terms of Christ on the cross, of in-decision.

Consider, for example, Tony Blair's confession in cognate terms, to the TV talk show host Michael Parkinson, of his own 'crisis' of decision over Iraq: 'Of course, you struggle with your own conscience about it ... and it's one of these situations that, I suppose, very few people ever find themselves in ... In the end, there is a judgement that, I think if you have faith about these things, you realise that judgement is made by other people ... and if you believe in God, it's made by God as well.'[4] The violent oscillation between tyrant and martyr thus introduces the plotter or intriguer; think of the Kissingers and Alastair Cambells of our world (Weber 1992: 17). In the theatrical space thus opened by the dislocation of the action and of the subject, and in the confusion that results, the sovereignty of the tyrant is replaced by the artifice of the plotter: 'In contrast to the spasmodic chronological progression of tragedy, the *Trauerspiel* takes place in a spatial continuum, which one might describe as choreo-graphic' (Benjamin quoted in Weber 1992: 16). Politically, we still inhabit the problem space of the baroque. Or, rather, we still inhabit the political temporality – that of factical finitude – whose problematisation of politics, government and rule finds its expression in the baroque character of modern politics, most especially its politics of security. Better to say, perhaps, since no doubt many other experiences of time are lived out in the midst of this modern conception of time, our exposure to apparatuses of rule is largely dictated by the operationalisation of this time politically. Thus, Weber concludes:

> The amoral calculatedness of the plotter contrasts radically with the attitudes of both the tyrant and the martyr. For only the intriguer confronts a state of the world in which the exception has become the rule, and therefore in which universal principles – and be it the principle of the interruption of principle qua decision – can no longer be counted upon. The intriguer exploits mecha-nisms of human action as the result of forces over which there can be no ultimate control, but which can therefore be made the subject of probabilistic calculations.
>
> (Weber 1992: 16)

The constantly repeated drama of the rise and fall of the prince appeared to the dramatists of *Trauerspiel*, adds Weber, quoting Benjamin, 'less as a manifestation of morality than as the natural aspect of the course of history, essential in its permanence' (Weber 1992: 9). History has thus become 'a repetitive and ineluctable process of rise and fall ... without any representable possibility of either grace or salvation' (Weber 1992: 9). It poses a problem space comprising 'an encirclement of contradictions and antitheses from which there seems no escape, but only the "dizziness", the vertigo', that its spectacles elicit (Weber 1992: 8). Gone is the Christian eschatology of soteriological finitude. The baroque knows no eschatology. For that reason, it possesses no mechanism 'by which all earthly things are gathered together and exalted before being assigned to their end' (Benjamin cited in Weber 1992: 12). Its history is no longer salvation history, its time is no longer the soteriological finitude of Christian faith. Its political temporality is that of the factical finitude of the modern as an infinity of finite things. Since its politics is the infinite government of finite things, its political anthropology is directed towards the cultivation of an infinite servility to finite government among its political subjects.

No eschatology here, then, but a time nonetheless entirely pervaded by a sense of an ending, because it is, of course, a time premised upon the infinity of finite things. It is infinitely saturated with endings. But these endings are endings of the same sort. They are incapable, in fact, of transforming the order of things. Another way of putting this point is that the baroque problematisation of politics, government and rule is obsessed with the infinity of endings that the infinite coming and going of finite things describes. An infinity of finite things differs from eschatology, in the messianic sense of the term, in as much as the time to come is simply more of the current time, the infinite becoming of finite things, rather than their transformation. For these reasons, the sovereign and his courtiers also cling obsessively to this extraordinarily contrived world of artifice, whose existence they call 'real', as if it owed its manifestation to forces outwith its making. Since the transcendent has become immanently invested in the modern sovereign, no world is allowed other than the world of political spectacle that he or she is condemned performatively to act out.

The drama of German *Trauerspiel*, and so also the object of Benjamin's analysis, is, therefore, not historical life as such. As if there could ever be such a thing. It is the time and timing of historical life as it comes to be posed and represented by its age, that of factical rather than soteriological finitude. Behind Benjamin's political anthropology, therefore, lies this political temporality. It is the political temporality of factical finitude, suffused with a pathos for the loss of the soteriological time in opposition to which it arose, that gives rise to modern political anthropology such as that of Hobbes, and all other modern political theorists as well.

It is this political temporality which poses the problematisation of politics, government and rule to which sovereignty has long been offered as the favoured answer. But, as *Trauerspiel* indicates, it simultaneously sets the problem that sovereignty is supposed to resolve in such a way that the sovereign is in fact incapable of solving it. Mourning for the loss of the Christian God becomes mourning for the loss of the sovereign as well, in which the sovereign assumes the strategic figure of Man, a mere afterlife of the Christian God. Modernity's pathos thereby gets amplified and intensified. Which is in some sense rather peculiar, since a measure of radical incredulity has evidently always surrounded the modern artifices by which both God and Sovereign make their appearance in the world, not least in the Christian world. That is precisely why the imperial Church, and the imperial state alike, emerged. They were *dispositifs* of power relations whose very *raison d'être* was to police, regulate and command the operation of the artifices by means of which God, and later Man, were said to order the world.

The factical destiny of the prince does not merely imply the rise and fall of an individual figure, however, but, more significantly, '*the dislocation of sovereignty as such*' (Weber 1992: 9). If concrete political form is given to finitude via sovereignty, then the predominant response offered to the problematisation of politics, government and rule in the modern period is always already deficient. The 'tragedy' of modern sovereignty consequently arises in the unbridgeable gulf between the sovereign's infinite responsibility to decide, in seeking to exercise the infinity of finite government that has devolved upon him or her, and their manifest inability to discharge the very task of which he or she is, in fact, comprised: 'However highly he is enthroned over subject and state [the sovereign's] status is confined to the world of creation; he is the lord of creatures, but he remains a creature' (Benjamin quoted in Weber, 1992: 13). Albeit the three figures of tyrant, martyr and plotter may pose as if they might be, they are in fact precisely not subjects independent of time, acting in time to command time. Time, in the form of factical finitude, is their staging. Creatures of their time, they are time's actors, posturing as its masters.

It is, then, clear that the dilemma of the sovereign in baroque drama is also and above all 'that of the [political] subject as such' (Weber 1992: 16). This modern political subject only appears to be determined by its 'head' – that is, 'by its consciousness, its intentions'. Even in its reasoning it is, instead, determined 'by forces that are independent of it, that buffet and drive it from one extreme to another' (Weber 1992: 16). Recall Bush and Blair again. They were driven by the 'tempestuous affects' to which they were subject, and over which they had little control (Benjamin quoted in Weber 1992: 16). 'Their actions are not determined by thought, but by changing physical impulses' (Benjamin quoted in Weber 1992: 16). The baroque gives expression to the ways in which politics, government and rule are problematised under the temporal conditions set by factical rather than soteriological finitude. It is thus the political temporality of the baroque, its factical finitude, that supplies the very animating force that institutes the figures that comprise its political anthropology, whose typology Benjamin explores in *The Origin of German Tragic Drama*.

What Benjamin and Weber fail to observe, however, is that, while Christian eschatology has disappeared, there is a figure left behind that continues to haunt the way in which the baroque stages and operationalises modern politics. This figure fails to make the cut of Benjamin's typology, and Weber's masterly exegesis of it. It is the figure of the Christian katechon. In the history of Christian salvation, the katechon's task was to restrain the chaos-inducing lawless-one stalking the earth, until God's Kingdom comes to pass and divine order prevails with the return of the Messiah. In the infinity of finite things that characterises factical finitude, it is not only transcendence, in the political figure of the sovereign, that has, however, been immanentised. So also, in the temporal figure of the infinity of finite things, has the eschaton. Where once the eschaton marked the end of the finite world and the transformative return of all creation back into the God's eternal presence, the eschaton is radically transformed as it is relocated by the advent of factical finitude, and through the ways in which the temporal continuum of the infinity of finite things is continually punctuated by the infinity of their finite ending. Nothing lasts. Everything comes and goes, in changing forms, but in a time unchanged by any transformation in the diverse and dynamic, contingent and unpredictable, ways in which they do appear. The eschatological break between the immanent and transcendent posed by the history of salvation, becomes a break that now continually interrupts the infinity of finite things. The continuum of immanence is thus constantly punctuated by an infinity of finite endings. The sovereign, later the state, assumes the role of katechon, and almost all modern international political theory is premised on the political necessity of the sovereign and/or the state assuming the role of the katechon to keep lawless *an-archē* at bay. It is OK that things come and go, so long as they as they come and go in ways ordered by the rules of truth and truths of rule of the modern political ordering of the appearance of things.

Into the deficiency of the sovereign, and therefore on to the stage that it fails to command, enter other formations of power, other rules of truth and truths of rule. Independent of the royal questions of power, as Foucault calls them, does not mean that these other *dispositifs* of truth and rule are of no interest to sovereign power relations, or do not themselves qualify, transform and newly empower as they undermine traditional accounts of sovereign power relations. Specifically, within the baroque field of formation and problematisation of modern politics arises biopolitics as well. Since fields of formation and problematisation are simultaneously also riven by surfaces of friction, and manifold points of application, crises of government, reversal of relations between truth and rule as well as dazzling spectacles of power, it is impossible to make this increasingly diverse theatre of power relations add up. Factical finitude acquires concrete political form in diverse and heterogeneous ways, as representation emancipates itself, baroquely, from the representing subject, and as the sign liberates itself, digitally, from the referent. The relation between the sign and the signifier is not guaranteed by the nature of the matter to which it was once supposed to refer, because, of course, there is a lack in the matter to which it was once said to refer. In such circumstances, the enemy, for example, becomes as unattainable as the Schmittean sovereign. Biopolitics is related to geopolitics in as much, therefore, as these are among the diverse modes through which factical finitude acquires concrete arabesques of palimpsestuous political form.

In short, had sovereignty not been so flawed and dislocated from its very conception and inception, perhaps it might have been sufficient to satisfy the problematisation of politics, government and rule posed by factical finitude. And perhaps not. After all, the advent of the infinity of finite things also witnessed the liberation of pluripotent forces of every description, including those attending to what Foucault called the conduct of conduct; or what, religiously and politically, the early baroque called 'decorum' (see Chapter 5, 'Pious Economies'; and Jonckheere 2012). In the temporality of factical finitude, sovereignty was in no position to monopolise warrant-granting powers over rules of truth and truths of rule released by the infinity of finite things. Instead, it always found itself in a market for strategic ideas concerning politics, government and rule precisely because its capacity to rule – to exercise infinite government over the infinity of finite things – was a conceit it could indulge only in obsessively contrived spectacles of sovereign power. Enter biopolitics, not as *dispositifs* summoned by sovereign power, but as a collection of diverse and changing *dispositifs* giving additional expression to the burgeoning rules of truth and truths of rule – forces and opportunities – released by the conditions of possibility and operativity of factical finitude.

Chapter 1, 'A political analytic of finitude: The infinity of finite government and rule', details the analytic just summarised here. Chapter 2, 'Biopolitics of security in the twenty-first century', summarises Foucault's account of biopolitics in order, then, to detail how, transforming the very ways in which the vital signs of life have been transformed by molecularisation and digitalisation, its very species being translated by these cognate processes into code, Life thus transformed transforms the operation of the biopolitics of Life. Chapter 3, 'Biopolitics, government and economy', does not address the economy or economies of biopolitics so much as explain how it is, according to Foucault, that the essential character of biopolitics is economic and emerges historically through the advent of such a thing called 'the economy', itself as much a governmental apparatus, of course, as it is a system of production, circulation and exchange, critically dependent upon species, in the form of money, as it is on species in the form of the properties of Life as species being. Life as species functions governmentally via species. The principle of utility that governs modern security politics governs the life of the modern economy as, itself, a form of species existence. When species escape, or vertiginously spin out of the regulatory mechanisms designed to fructify and capitalise their capacity to go forth and multiply – 'making live' – crises of economy immediately translate into crises of government just as crises of government immediately translate into crises of economy. Chapter 4, 'Underwriting security', elaborates the account of biopolitics offered in Chapters 2 and 3. It explores how insurance and finance capital paradigmatically illustrate the operation of a form of politics, government and rule whose referent objects of rule are the changing properties of species Life via the Life of species, especially that of money, through the now pervasive technology of risk.

Chapter 5, 'Pious economies', shifts the tone and historical register of what has been offered so far. It goes back to the advent of the modern age in the late seventeenth century and, in particular, to the Baroque German town of Halle in Brandenburg-Prussia. There, it explores the emerging new modern power plays of religion, politics and economy – staged through the complex intersections of the revised pastoral powers of evangelical Protestantism, inspired by Herman Francke, with the new powers of state formation and police inspired by the political theorist and jurist Christian Thomasius and his *decorumstreit with* Francke – as these were played out globally in places as far apart as North Germany, North America, Russia and the Coromandel coast of the Indian subcontinent. The chapter explores the religio-political nexus of the modern age as Christian churches became complicit with the developmental governance of Life. It brings the story up to date by addressing how, in the late twentieth and early twenty-first centuries, religions now classified governmentally as 'faith-based organisations', are themselves securely governmentalised via their recruitment into biopolitically inspired neo-liberal global development policies.

Chapter 6, 'Biopolitics of security, race and war', explains how Foucault understood the link between race and biopolitics. It proceeds to do what other chapters have done by exploring how, in addition to its explicitly racist inheritance from the earlier imperialisms of geopolitics, race comes to find algorithmic modes of biopolitical operationalisation in the digital age. Chapter 7, 'Governing terror: The state of emergency of biopolitical emergence', is the last substantive chapter. Earlier chapters explained how Life in the digital and molecular age came to be understood in terms of continuous complex emergence. Emergent Life – being-in-formation – continuously also poses a new security problematic, that of the infinity of becoming-dangerous to which it is given from the perspective of biopolitical governmental construal of it. Life as emergence is, therefore, Life as continuous emergency.

Just as the book begins with an account of how the modern sovereign consists in a task it is constitutively incapable of discharging, so also it ends with how contemporary biopolitics of security also consists now in a task that it is constitutively incapable of discharging. Emergent Life consists in a continuous emergency of emergence. You cannot bring an end to that emergency without bringing an end to Life. This would contradict the very mission of biopolitics. Contemporary biopolitics seems, therefore, committed to doing what the sovereign of *Trauerspiel* did, seeking the tyrannical powers required to discharge the infinite and thereby impossible task of which it is in fact finitely comprised. The powers that it seeks are not, however, those of katechontic restraint of which earlier sovereign politics was comprised. The vertiginous emergency of emergence that this inspires in modern government cannot, therefore, be allayed by restraining the lawless chaos that threatens the law. Something extraordinary happens instead. It seeks to incorporate the powers of chaos into its very katechontic function. Together with my colleague Julian Reid, I explored how this began to take place in the revision of military strategic discourse of military-technical revolution that swept through 'the liberal way of war' (2009) during the course of the last decade of the twentieth century, and the first decade of the twenty-first century. In the 'Conclusion' to this book I revisit the account of the katechon given by St Paul in his 'Letter to the Thessalonians' and the way in which Carl Schmitt dimly intuited and anticipated how the katechontic function of modern politics, government and rule begins to operate through 'acceleration' of the forces of chaos rather than their mere restraint. Some security.

Notes

1 There is a countervailing positivity to this story. It concerns how the discovery of 'The New World', in so profoundly impacting the theologico-political and cultural imagination of the old, had upon the advent of the baroque as well. See especially Pagden (1993).

2 Part of the point that Turgot was making in the epigraph that heads this book. (The quotation comes from Benjamin and is cited in Kapczynski 2005.)
3 For a detailed and comprehensive account of the changing nature of the subjectivity of spectatorship, and of the very plurality of the category 'spectators', see for example, Harris (2014).
4 <http://news.bbc.co.uk/2/hi/4772142.stm>, accessed 3 August 2014.

Bibliography

Benjamin, Walter (2009 [1925]) *The Origin of German Tragic Drama* (London: Verso).

Blumenberg, Hans (1985) *The Legitimacy of the Modern Age* (Boston, MA: MIT Press).

Butler, Justin (2012) 'Re-reading the Grácian Self in a Self-Made World', *Hispanic Issues Online Debates*, 4: 1–23.

Deleuze, Gilles (1993) *The Fold: Leibniz and the Baroque* (London: Athlone).

Dillon, Michael and Reid, Julian (2009) *The Liberal Way of War: Killing to Make Life* (Abingdon: Routledge).

Gal, Offer and Chen-Morris, Raz (2013a) *Baroque Science* (Chicago: University of Chicago Press).

Gal, Offer and Chen-Morris, Raz (eds) (2013b) *Science in the Age of the Baroque* (Dordrecht: Springer Press).

Harris, Joseph (2014) *Inventing the Spectator: Subjectivity and the Theatrical Experience* (Oxford: Oxford University Press).

Hobbes, Thomas (1960) *Leviathan*, ed. Michael Oakeshott (Oxford: Blackwell).

Jonckheere, Koenraad (2012) *Antwerp Art after Iconoclasm: Experiments in Decorum, 1566–1585* (New Haven: Yale University Press).

Kahn, Victoria (2003) 'Hamlet or Hecuba: Carl Schmitt's Decision', *Representations*, 83(1): 67–96.

Kapczynski, Amy (2005) 'Historicism, Progression and the Redemptive Constitution', *Cardozo Law Review*, 26(3): 1041–1117.

Maravall, Jose Antonio (1986) *Culture of the Baroque: Analysis of a Historical Structure* (Minneapolis: University of Minnesota Press).

Pagden, Anthony (1993) *European Encounters with the New World* (New Haven: Yale University Press).

Weber, Samuel (1992) 'Taking Exception to Decision: Walter Benjamin and Carl Schmitt', *Diacritics*, 22: 5–18.

1

A POLITICAL ANALYTIC OF FINITUDE

The infinity of finite government and rule

> So, naturalists observe, a flea
> Hath smaller fleas that on him prey;
> And these have smaller fleas to bite 'em;
> And so proceed *ad infinitum*.
>
> (Swift 1958)[1]

> ∞ ...[2]

Introduction

> the *usus facti* in which Franciscan poverty is manifested is identified without remainder with the renunciation of ownership and not ... with an intrinsic characteristic of use itself, the *usus pauper* (poor use).
>
> (Agamben 2013: 127)

The thought always escapes the finite horizons of the author. The thinker cannot therefore own the thought. Usufruct rather than ownership thus inspires the ethic of reading that informs this book. Drawing its inspiration, in addition, from 'continental philosophy' more generally, the book uses Michel Foucault, puts him to work in pursuit of one of the tasks that he set us. In this instance that task is not simply the history of the present, but accounting for the overwhelming presence of security in the political history of the present.

In respect of government and security, the present comes down to us, first, in terms of the alliance between the state and security. Later, with Foucault and especially the lecture series *Society Must Be Defended* (2003) and *Security, Territory,*

Population (2007), it comes down to us in terms of governmentality and security, specifically governmental biopolitics in which security and government revolve around the properties of species existence, rather than those of sovereignty. The emergence of 'population' was one early, positivistic, expression of the governmental biopolitics of security. It still remains an important point of application for governmental biopolitics. However, since the digital and molecular revolutions of the twentieth and twenty-first centuries, the preoccupations with virtual life, artificial intelligence, risk, complex adaptive emergence, 'eventalness', complex emergencies and resilience have also become important.

Society Must Be Defended and *Security, Territory, Population* summarily also remind us of the following additional and related points: first, that all forms of rule require to be operationalised governmentally; second, that modern governmental apparatuses, such as *polizeiwissenschaft* and discipline, arose simultaneously with the emergence of modern power relations, and in response to a generic crisis in the 'conduct of conduct' precipitated by the fragmentation of Christianity, as well as the discovery of the New World (Pagden 1993) and the rise of global colonisation, modern economies and modern state sovereignty; third, that a complex relation obtains between modern government and the modern state, such that Foucault suggested that the modern state has been subject to comprehensive governmentalisation; fourth, that modern government and rule is, thus, characterised by the coexistence of many diverse and heterogeneous power relations; and, finally, that the modern biopoliticisation of government has introduced a novel security/government complex revolving around the positivistically established properties of species being, one that threatens to colonise all other modern forms of politics, government and rule throughout the North Atlantic basin. A *mentalité du gouvernement* finding expression in a *mentalité de sécurité* deriving initially from adopting the sovereign's gaze – but now also associated with the widespread biopoliticisation of modern politics, government and rule – threatens to displace all thought of modern politics, not least democratic politics.

This opening chapter is especially concerned with understanding how the modern imperative to securitise finds its historical roots, and many of its distinguishing analytical features, deep within modernity's temporal enframing of finitude. Its conclusion is that biopolitics of security was one of the ways in which concrete political form was given to the factical finitude that succeeded the soteriological finitude which once temporally enframed western problematisations of politics, government and rule. In contrast to almost all discussions of Foucault and security which substantially confine their accounts to *Society Must Be Defended* and *Security, Territory, Population*, however, this book of essays on biopolitics of security in the twenty-first century is more deeply indebted to Foucault's magnum opus *The Order of Things* (1989). It is there that Foucault explains how the epistemic shift to modern finitude transformed modern forms of knowing. Just as Foucault thereby offered an epistemic analytic of modern finitude, this book offers a complementary political analytic of modern finitude, which finitude it calls 'factical finitude' (Raffoul *et al.* 2002,

2008; Stambaugh 1992). It explores how this modern factical finitude – defined by an infinity of finite things – sets the problem space for modern politics, government and rule. Doing so, it not only accounts for the centrality of security in general, and biopolitics of security in particular, to modern discourses and practices of politics, but for their baroque and deeply katechontic character as well.

For, if (a) the baroque is less a period than a problem space, as its contemporary students persuasively argue (Maravall 1986; Lambert 2008; Egginton 2010); and if, as they also argue, (b) the problem space arose directly out of, and in response to the political as well as the aesthetic demands posed by the very crisis in the conduct of conduct to which Foucault refers, and which gave birth to early modern Europe; then (c) the enframing in terms of factical finitude set the temporal horizons of the baroque as well. Put differently, factical finitude found its very mode of expression as well as its operationalisation in the tones, both political and aesthetic, of the baroque.

Just as modern politics, government and rule are thereby problematised by calculations to infinity, epistemically introduced in response to factical finitude as a means of bringing the infinity of finite things into the light of finite positivistic knowing (Campe 2012; Clegg 2003), so, also, are modern government and rule, and their burgeoning security apparatuses, equally characterised by calculations to infinity. For it is through calculations to infinity that the positivities of finite entities to be secured are specified, so that they can in fact be secured, not least in respect of the infinity of finite dangers that they necessarily also present; you cannot secure any finite thing without knowing something positivistically about its finite properties.

Contemporary risk calculations of security provide one classic example, and, in addition, they do so for entities that are not given preformed and comprised of fixed properties, but exist contingently; consequently, they have to be secured probabilistically because they are temporal finite entities whose properties are a function of complex adaptation and change through time with other entities, as well as in response, where living things are concerned, to the operation of their molecular structure. Such is the factical account of finitudinal Life that now dominates Life discourses in the twenty-first century. For, under the terms of factical finitude, Life is an entity whose finite course is as much contingently determined as its very event is classified as a cosmic accident (Kauffman 1993, 1995; see also Koyré 1957; Harries 2001). This is as much an ontological as it is a biological account of Life.[3]

It is for these reasons that modern government and rule have become a ceaseless katechontic task, infinitely seeking to secure the very finitude of which they are comprised against the indefinite ending that assuredly defines and awaits them. Unlike soteriological finitude, factical finitude is not, however, governed by reference to some final end of days. As Foucault observed, and many others like Reinhardt Koselleck (2000 [1988]) have also concurred, there is no end, as such, to the problematic of modern government and rule: 'There will always be governments, the state will always be there, and there is no hope of having done with it' (Foucault 2007: 355–6). Here, existence persists *ad infinitum*, repetition without termination, rather than *sub specie aeternitatis* describing an ever more baroque spiral of security technologies and specular forms (Rotman 1993a, 1993b, 2000; Kaplan and Kaplan 2003; Lambert 2008). There is, thus, no Christian end of days for factical finitude, only an infinite succession of finite entities whose days are always numbered, among them not only finite human beings but also finite regimes of government and rule, that also assuredly come and go within its infinite compass. The problematic of modern politics government and rule, *a fortiori* the problematisation of modern *dispositifs de sécurité*, is thus enframed by a factical rather than a soteriological finitude. It is one that proceeds, politically as well as epistemically, through the securing and governing positivistically of an infinity of finite things, and their processes of becoming, rather than the governing and securing spiritually, underwritten by the promise of life everlasting.

Finitude

> Maybe we could say that *raison d'état* basically posited as the primary implacable law of both modern governmentality and historical science that *man henceforth has to live in an indefinite time.*
>
> (Foucault 2007: 355; emphasis added)

Infinity is not a synonym for eternal life. For one thing it is entirely indifferent to the fate of human kind. For another it contains no promise of the conclusion of time; if indeed there is to be a conclusion to time, or a conclusion to time that has any current bearing on the human condition. Precisely because factical finitude poses no extra historical standards for political judgement, when confronting the rules of truth and truths of rule of politics, government and rule, factically, human beings do not lose the faculty of judgement. '[O]n the contrary, they acquire it.'[4]

Christian political theology is in many ways the most forensic analysis of the soteriological problematisation of politics, government and rule, in terms of sovereign power, that we have available to us. Postulate the existence, or the need to have, a single homogeneous creative agency in order to institute goodness and order throughout the world, then political theology tells you what happens next. Political theology is political theory. It teaches that the entailments of sovereignty entail hierarchy, inequality and a principled hostility to heterogeneity. More than that, it also teaches that sovereignty consists in a civil conflict (*stasis*), one in which a drive to homogeneity continuously seeks to overcome heterogeneity but fails to do so. Posing power over death as the principle of formation that will institute peace – the sacrificial redemption of sin, or the instrumentalisation of war – it fails instrumentally to command the very power it is said to consist in.

As Victoria Kahn has also argued recently, 'why should we assume that religious metaphor is a subset of political theology, understood as the notion that there is some transcendental ground of political authority? Why not assume the exact contrary, that political theology is a subset of metaphor, that is, of purely human ways of understanding the world and negotiating its political conflicts?' (Kahn 2014: 7).

This, she goes on to say, 'is the view of Benjamin, Kantorowicz and Cassirer, as well as the view of Machiavelli, Shakespeare and Spinoza' (Kahn 2014: 7). It is also mine.

I do, however, depart from those, like Kahn, who more generally subscribe to the naive modern view, advanced initially by Vico and by Hobbes, that we can know only what we have made or constructed ourselves – *verum et factum convertuntur*. If only it were that easy. For, manifestly, we do not, and we cannot. Our very own artefacts exceed the understanding that went into the making of them. Our projects have a life of their own, baroque systems of formation and re-formation that far exceed the scientific, technical and aesthetic artifice, as well as semantic fields of reference, that went into their making. We continually seek to catch up with those lives in the effort to bring them back under our control, to end them, or direct them to different ends. What applies to our institutions and practices applies equally, of course, to ourselves. In as much as we may also make ourselves, factically, in the very enacting of ourselves we loosen the ties to the inherited selves with which we have been gifted, or damned.

Contra Hobbes especially, then, the political truth of the political artefacts that we make remains opaque even to those in government that operate them. The inseparability of clarity from opacity is another classic baroque trope (Deleuze in Egginton 2010: 134, n. 7). The fundamental opacity of modern political truth to modern political truth makers, as much as those governed by such truths, does not however undermine the point about metaphor and political theology made by the Renaissance and early modern artists and thinkers that Kahn endorses. It reinforces the demand for it, and the obligations that go along with it. If factical finitude poses no extra historical standards for political judgement, when confronting rules of truth and truths of rule of modern politics, government and rule – in their very opacity to those who inhabit them, and the marvels by which they contrive to turn us into spectators more governable because we have been induced to fall for them – I say again we do not lose the faculty of political judgement factically; on the contrary, we are presented with an opportunity to temper and refine as we acquire it.

As Maravall (1986), Lambert (2008) and Egginton (2010), among other students of the baroque and neo-baroque, note, the baroque is a space of problematisation not a mere period of artistic expression. It is one open to counter-strategies of intervention as well. As the latter two authors, in particular, note, the baroque also poses possibilities subversive politically of itself, if it is, for example, a classical baroque political trope to work on the affects of imagination via the spectacles of rule – Hobbesian artifice for example. The purpose is to render us more governable material. For how else is the infinite government of finite things to be exercised? If factical finitude poses no extra historical standards for political judgement, we can equally also, however, exercise our imagination so that we do not fall for the marvels upon which modern baroque politics so regularly relies the better to render us more governable. It takes little tempering of the imagination, in fact, to imagine that the offerings of the baroque spectacles of modern politics are the product of artifice; and not least those of its politics of security.

Facticity thus begins but it does not end in factuality, not even a factuality that is clearly of our own making. To accept that there is an excess of factical being over its appearance, for example, concedes nothing to theists and traditional transcendentalists. On the contrary, it is what makes the task of factical freedom a difficult freedom whose obligations and responsibilities cannot be projected on to some transcendental outside. The practice of regularly doing so is itself a standard trope of politics, government and rule, a hackneyed, if nonetheless effective, baroque political manoeuvre as common to late modern liberal imperialism as it was to early modern Spanish state building (Maravall 1986).

Neither is this a solipsistic teaching concerning human being. It cannot be. For, as selves, factically, we do not simply inhabit the improbable openness of factical finitude, we are direct expressions of it. In the process of enacting such existence, we value and make it meaningful, not once, not once and for all, but continuously and with revision. In Arendt's terms, factical freedom is thus, 'to be confronted anew, without the religious trust in a scared beginning and without the protection of traditional and therefore self-evident standards of behaviour, by the elementary problems of human living together' (Arendt in Kahn 2014: 181). Christian transcendence is not so much reoccupied as a space of problematisation, it is simply replaced by factical finitude. I make no claim for the superiority of that space. I claim only that it is a radically different space of problematisation, one giving its own expression to the horrors of human existence, but equally also calling upon the greatest powers of human generosity, endeavour and inventiveness.

Neither is the position advanced here a secularist one. The secular arises as a category of religion. Secularisation is an expression of the soteriological rather than factically finite problematisation of politics, government and rule. Secularism has, of course, been a persistent and significant problem *for* political modernity, but it is not the problem *of* political modernity. It is, however, the problem *of* revealed religion and, in the form of the Christian tradition, it has been so for millennia. Thus the problematic of secularisation – the submission of religious to temporal government – was a defining issue for the Christian theologico-political tradition going back to Jesus' enigmatic teaching to 'render unto Caesar what is Caesar's, and to God what is God's', as well of course to Constantine and Eusebius.[5] Hence, Christianity's political theology concerns the politics, government and rule posed by faith in the incarnate, as well thereby also of the triune, Christian God. It does not concern the politics, government and rule of the factically free.

Factical finitude is, however, a space of problematisation that continues to exist in heterogeneous relation with the soteriological space of problematisation offered by revealed religion. Revealed religion retains a massive influence in and over the world, in societies that thought they were secular as much as in societies that expressly present themselves as religious. But, as the globalising history of the twentieth and twenty-first centuries has made perfectly evident, there is no such thing as a purely secular or a purely religious society. If modern power relations are diverse and heterogeneous, so also are its spaces of problematisation.

Much less, also, is the position explored here a post-secularist one. Just as it makes no claims concerning the decline of religion in the public and private lives of western societies – it would be foolish to do so – neither is it concerned with claims concerning their revival. I don't think religion ever went away, and, therefore, remain unimpressed with arguments concerning its return. The modern has always been a religio-political complex, which is what my argument concerning the persistence of soteriological and factically finite problematisations of politics, government and rule within it maintains. But it is the factically finite that distinguishes the project of modernisation. My position is therefore founded, instead, upon a radical incredulity towards the narrative teachings of revealed religion. It shares that radical incredulity with the move towards the enframing of finitude as factical, but does so without embracing all of the many diverse ways in which facticity finds its expression, including that of the baroque, or sharing all the philosophical positions of those modern philosophers who have championed it – Heidegger, for example. For me, factical finitude continues to remain an open question. How else could it be?

When Schmitt says, 'All significant concepts of the modern theory of the state are secularised theological concepts' (Schmitt 2005: 36) we should, therefore, take him at his word. Yes they are, but only insofar as the modern state is concerned. And, with a significant difference also, because, as the epigraph from Foucault makes clear, the modern state exists within – indeed as it emerged through struggle to escape the discourses and structures of medieval politics, government and rule it actively contributed to the inauguration of – the indefinite time of modern finitude (Cavanaugh 2002; Strayer 2005; Dupré 1993). But it is Schmitt's theologico-political problematisation of rule, without the *theos*, that makes him such a characteristically baroque political thinker. Or, one might more precisely say, instead, a political thinker steeped in the problem space of the baroque.

Thus, while the modern state inherited a theologico-political problematisation of politics, government and rule via sovereignty (Lefort 1988; Gillespie 2008), its operationalisation of that problematic now takes place within a temporal order quite different from that which gave rise to it. If sovereignty's theologico-political problematisation of politics, government and rule is a direct expression of Christianity's soteriological order of faith, its modern articulation now takes place within a time conceived as the factical infinity of finite things. Human finitude finds quite different expression in these rival accounts of time. Each temporality also offers a different spatio-temporal enframing in which politics, government and rule, along with security, are problematised in fundamentally different ways. Here, as another student of the baroque observed:

> The old symbolic cast of mind, with its assumptions of an ordered and hierarchical cosmos, remains operative until well into the second half of the seventeenth century, but an irritable doubt as to the precise relationship between seen and unseen worlds informs the Baroque, in both its typical works and its masterpieces.
>
> (Warnke, quoted in Egginton 2010: 131, n. 5)

From the perspective of soteriological finitude, human being is *ens creatum*. It occupies a privileged place in God's creation. It lives out its finite existence in the promise of life everlasting. Its strategic figure is Man, a creature different from all other creatures because, by grace of God, it is endowed with a divine spark. Soteriologically speaking, there is a why and a how, but, despite revelation, the why and the how ultimately retain their mystery. Christian politics has thus been a radically opaque politics of guilt, sin and salvation, as much as a politics of conditional forgiveness and love. It is a divine security politics lived out in pursuit of everlasting life (the security of salvation) according to the plan of an inscrutable if beneficent deity (Delumeau 1990; Kaufman 1990; Moore 1987).

Conversely, from the perspective of modern factical finitude, life as such is a mere accident of cosmic contingency, and human existence is an expression of its evolutionary dynamics. Both, cosmic contingency and evolutionary dynamics, are knowable, positivistically, but are ultimately also unintelligible. Factically, there is no why, there is only how. Man is the strategic figure of soteriological finitude. Life has become the strategic figure of factical finitude; human existence is just another expression of Life and the vast infinity of finite things in which Life quite contingently arises, and finds its contingently finite expressions (Monod 1997). Life is, in short, a factical phenomenon. Man, conversely, is a soteriological phenomenon. The after-life of the Christian God, Man is a zombie-like extension into a time long released from the governing assumption of God's ubiquitous presence. These figures happen also to coexist in modern times, whose power relations are consequently diverse and heterogeneous as well as palimpsestuous (Genette 1997; S. Dillon 2007). Its security problematics, therefore, overlie and complicate the reading and interpretation of one another.

It is only if you ignore the political analytic of finitude, and assume with Schmitt that the defining question of politics remains that of the theologico-political concept of sovereign power (now vested in the state by reference to the metaphysics of popular rather than divine sovereignty), that you can equate the modern problematisation of politics, government and rule with sovereignty alone; and thereby also with a political theology now confronted by factical finitude's radical incredulity towards religious narratives, liturgies, *oikonomia* and power relations. Schmitt advances his claim on the basis of soteriological assumptions concerning time and human finitude derived from the God of revealed religion. With all that this entails. Entails not only in respect of specifying the nature of God, but the governmental entailments, also, of the divine economy or *oikonomia* of Christianity's three persons in one God, and of how that God's will becomes actualised throughout all creation (Agamben 2011). For such issues – posed classically, persistently and most dangerously to the Christian message by Gnosticism (Jonas 1952, 2001) – may fatally compromise the Christian teachings of the indivisibility, goodness and sovereignty of the Christian God.

In the event, the primary entailment of divine governance on earth, continually rivalled also, however, by the complex messianic inheritance of Christian belief, became katechontic. Schmitt's apparent loss of faith in Christianity's messianic message – complemented by his deep interest in baroque political thinkers like Malebranche (1638–1715) – leaves the jurist and political theorist in him with nothing but the burnt-out figure of the katechon. For what remains of a world left bereft by the deeply compromised redemptive promise of a *deus absconditus* but radical evil and *ressentiment*? And where, but in the figure of Schmitt's katechon, are *ressentiment* and radical evil concentrated most toxically than in his definition of political decision-making as the nihilistic drawing of the friend/enemy distinction?[6] Thus, in its loss of onto-theological faith, does Schmitt's baroque sovereignty offer concrete nihilistic form to the problematics of politics, government and rule posed by the waning of the rules of truth and truths of rule of soteriological finitude, and the advent of those posited and posed by factical finitude.

In further interrogating the onto-theological analytic entailments of sovereignty it is helpful to compare sovereignty and slavery. Katechontic, the sovereign is the rule of the one. Slavery is the rule of the master, as the one. It, too, is katechontic since slave revolt is the most inherently dangerous threat to slave society. Just as there is no master/slave relation without the equality that enables the slave to understand and operationalise the order of the master (Rancière 1999), so also there is no order of the one (sovereignty) without the heterogeneity against which it defines itself and institutes sovereign rule. Equality does not precede or succeed slavery. Paradoxical as it may sound, equality is integral to slavery (Rancière 1999). Heterogeneity does not precede or succeed homogeneity. Paradoxical as it may sound, heterogeneity is integral also to the rule of the one. Neither slavery nor sovereignty can in fact be thought, much less operationalised, without equality on the one hand, and heterogeneity on the other. Slavery – the rule of the master – is the denial of equality. Homogeneity – the rule of the one – is the denial of heterogeneity. It is the very denial that acknowledges the existence of equality in slavery and heterogeneity in sovereignty. It is the operationalisation of that denial which institutes slave and sovereign regimes alike.

Homogeneity, presupposed by the rule of the one, also requires hierarchy, whereas heterogeneity, denied by the homogeneity and hierarchy of the rule of the one, expresses equality. Not equality before God or the Law but the equality of heterogeneous singularities. For heterogeneous difference arises from the singularity of each and every one (Clark 2005). Such singularity does not consist in a determinable property. It is inoperative (in the sense that Nancy and Agamben use that term: Agamben 2011; Nancy 1991). Neither does it consist in the good. It is good for nothing (Dillon and Reid 2009). Its freedom, politically, is factical freedom. That facticity lies in the operational capacity – of the magnitude 'that escapes any ordinary measurement' which distinguishes the singularly heterogeneous and heterogeneously singular – to claim equality (Rancière 1999: 15). Its political discourse is correspondingly polyphonous. Presupposing homogeneity, the rule of the one is the rule of the master. Instituted against heterogeneity, it is similarly also the denial of the equality of heterogeneous singularities.

Thus slavery presupposes the equality of the slave, even as equality is ideologically denied in order to institute a slave regime. Similarly, sovereignty presupposes heterogeneity, even as heterogeneity is ideologically condemned as radical evil, or domesticated as plurality, when instituting a sovereign regime. Slavery requires the whip. Sovereignty requires an apparatus of peace-making organised according to the grammar of war, the one that it wages against heterogeneity. That apparatus is a security apparatus. The discourse of security is its discourse of war articulated as the discourse of peace.

Agamben's analysis of the *oikonomia* of the Christian Church (2011) is thus deficient in this one crucial respect. Since its very inception in the combat myths concerning the war in heaven, the *oikonomia* of the Christian Church has always been conceived and operationalised as a war economy. The war in heaven extended via Lucifer, and man's fallen nature, into an endless war against radical evil does not cease even with the projected end of the world. For, whereas the army of angels and saints waging it will presumably be stood down for ceremonial duties once the Kingdom has finally arrived, the policing of hell thereafter nonetheless calls for an eternity of low intensity operation (Auffarth and Stuckenbruck 2004; Reed 2005; Keck 1998; Peterson 1964; Forsyth 1987).

Consequently, security has never been a mere service that a sovereign supplies to its people. The very metaphysics of sovereign rule, grounded in the homogeneity of the sovereign as well as the hierarchically organised mechanics of sovereign government, are founded in the originary requirement to secure the sovereign against the heterogeneity nonetheless integral to it, that threatens to assail it. This is where its katechontic enterprise is originally rooted. But it is deeply complicated also by the compound fracture that obtains between metaphysics and mechanics, reigning and governing, that comprise sovereignty itself.[7] Thus in the tradition of Christian political theology, and indeed much of the onto-theologisation of Western political theory, albeit indebted also to Aristotle, the metaphysics of rule deals in timeless principles because it presumes to deal in unchanging forms, whereas its mechanics of rule deals in contingent mechanisms because it addresses changing matter. Thus, the onto-theologico-political metaphysics of sovereign rule proclaims rule's essential mode of being, while the mechanics of sovereign rule specifies its changing modes of operationalisation. These are heterogeneous tasks. Integral to the very structure of sovereign government and rule, their intersection constitutes an indefinite surface of friction, comprised of an infinity of finite points of application endlessly posing the question of how they are to be allied, or, at least, contingently reconciled with one another.

The metaphysics of sovereign rule thus claims to be independent of time and space. The actualisation of that rule is, however, entirely dependent upon the mutability of governable material and the changing spatio-temporal exigencies in which it arises. Thus the economies of sovereign rule, the governmental technologies of *oikonomia*, concern the mechanics rather than the metaphysics of rule: the apparatuses rather than the principles, the how rather than the what. In the process, the mechanics of rule necessarily take their cue from the contingent properties of the governable material that rule seeks to govern, and the changing circumstances in which such government takes place (Foucault 2007; Agamben 2011).

Mechanics are as essential an element of sovereign *dispositifs de sécurité* as metaphysics. Sovereigns have always required their governments, just as governments have always required their sovereigns. The one secures the other, operationalises rule or legitimises operations.

Combined, as political metaphysics and governmental mechanics, sovereignty and government proclaim the rule of the one. And yet they cannot make it so, since there is always already more than 'one' at play within the sovereign itself. This more than one is no plurality. Like metaphysics and mechanics, sovereigns and governments are heterogeneous entities whose unity can never be established. It can only ever be simulated. And it is simulated through the manifold baroque spectacles of power (Debord 1994; Lambert 2008; Egginton 2010). That is why spectacle is no mere ideological false consciousness cynically deployed by power holders. Such it may also be. But, analytically, it is also integral to the rule of the one. Otherwise there would be no self-presentation of the one. However bloody, however technological and intrusive, geopolitically as well as biopolitically, security apparatuses may be, they are unable to function effectively without their essential measure of baroque spectacle simulating a unity that must be glorified because it is unattainable (Agamben 2011; Lambert 2008; Egginton 2010).

Combined, they also claim to secure the infinite regression of rule against some final sovereign ground. And yet again they cannot realise this ambition either, since the very compound fracture of which sovereignty is always already comprised poses an endless and irresolvable problematic of rule; that of the relation of metaphysics to mechanics, of the government to the sovereign, or of the Christian Church's *oikonomia* of religious government to belief in the divinity of Jesus Christ.

The distinction to be drawn between the metaphysics and mechanics of rule – between the maker and the made, and between the willer and the will – has at all costs therefore to be secured, as much through continuous simulation as through endless technological invention, if the very conception of sovereignty itself is to be sustained. The one must always be held distinct from, and held in hierarchical relation to, the other. The job of the sovereign is to rule or reign. The task of the government is to operationalise that rule. Sovereign accounts of power must, then, explain how the maker and the making – the metaphysics and the mechanics of rule – are to be related, for the one always threatens to diverge from, or become fatally conflated with, the other. Governments seek to become sovereigns, and forget to govern. Sovereigns seek to govern but, overwhelmed by the infinite government of finite things, forget to reign. Governments fail to implement sovereign will. Sovereigns are often found to be deficient in will, or its strategic direction. In the one case the sovereign intervenes to assert his will. In the other the government intervenes to invoke sovereign will, or improve its strategic management. As for 'the people' – a tragic, abused and martyred 'body' if ever there was one – having faith in their sovereign and love for the institution, they give thanks for the peace and prosperity it secures, or petition their sovereign over his government to redress their abject condition. In the final analysis, they may revolt.

In some respects this is the story that Agamben's *The Kingdom and the Glory* (2011) rehearses. The key point to be added here, however, is this. That relationship has to be explained and secured in ways that continuously support and reproduce the distinction that gives rise to it. It does not resolve, but must continually secure, the continuous reinstitution and ritual display of the unity of hierarchical harmony between the maker and the mechanics of making. If it did not, then the sovereign apparatus itself would have to be abandoned. One might even say that sovereigntist responses to the problematisation of politics, government and rule are nothing but the securing of what they must presuppose, the oneness of government and rule. Securing metaphysics' sway over mechanics, while suppressing heterogeneity's claim to equality, poses the very securitising problem space of sovereign politics, government and rule. Baroque spectacle and glorification are integral to securing this hierarchically organised unity since none such can be made to appear without them. So much is evident throughout the changing discursive history and operational practices of sovereign power.

Such now happens under the very different logics and dynamics posed by factical rather than soteriological finitude, however, and by reference to sovereignty's own continuously threatened end of days, rather than that proclaimed by Christian faith. 'The King is dead, long live the King!' But the Eucharistic doctrine of the King's Two Bodies (Kantorowicz 1946, 1998; and, Schwarz 2008) no longer serves to explain how or why, other than in response to the infinite drive to secure the impossibility of finite self-preservation. Thus, finally, according to Schmitt, but many others also concur, modern liberal rule is derelict because it represents the triumph of the contingent mechanisms over the essential metaphysics of rule: liberal governmentality over sovereign decision, the contingent administration of things over the expression of their essential political substance. It is economy without rule, dominion without reign. Governing to govern, it secures to secure, serially betraying its principles of representative and accountable government, respecting the privacy and individual freedom of its subjects.

In modern times, therefore, both the sovereign and governmental problematic of politics, government and rule are enframed by factical rather than soteriological finitude. Both, modern political sovereignty and liberal governmentality, now increasingly define their task of politics, government and rule in relation to the infinity of finite things. Security is less an epiphenomenon of sovereignty than its very *raison d'être*. Biopolitical security is less an epiphenomenon of liberal governmentality than its very *raison d'être*, infinitely engaged in the reduction of Life to positivistically ascertainable features that will secure its infinite governance. Together, sovereign rule and biopolitical governance threaten to exhaust the field of formation of which modern politics enframed by factical finitude might alternatively be comprised. How, is a question the dim outlines of which are only beginning to emerge.

Whereas sovereignty is, therefore, a security apparatus, generally legitimised as a protection racket but preoccupied with securing itself against its own finitude, government is the infinite securing of the violent emergence and passing away of an infinity of finite things. Thus everything is dangerous in its factically finite, and infinitely contingent, heterogeneity to sovereignty, whereas everything is dangerous to biopolitical governance in the very emergency of its continuous emergence. In reducing politics to the government and rule of sovereign security, geopolitics wars against heterogeneity and equality. In reducing life to what can be positivistically known about it, biopolitics reduces life to the Life of species existence. Thus does each give concrete political form to factical finitude, while betraying the political challenges of freedom, equality and ethical conduct that may otherwise issue from it.

Threatening now to overwhelm all other modern political claims, ambitions and promises, the infinite securing of finite things turns sovereigns and governments alike into katechontic *dispositifs* committed to infinite securitisation in indefinite time. This very apocalyptic responsibility has come to characterise the baroque tragedies of modern sovereigns (Benjamin 2009 [1925]; Koselleck 2000 [1988]) as much as the Kafkaesque security politics of modern governments. If the one is first interrogated in the *Trauerspiel* of the baroque the other is exemplified in the will to total surveillance disclosed by the whistle-blowing of Assange, Snowden and Manning, together with the governmental rage and *ressentiment* that it has elicited (Leigh *et al*. 2013; Harding 2014).

The sovereign machine must therefore always secure itself against the threat posed by rebellion in the name of a heterogeneity that claims an equality that fundamentally subverts sovereignty, or by a *coup d'état* that seeks to institute a truer or more effective sovereign (Foucault 2007; Rancière 1999). *Coup d'état* never, however, threatens the structure of sovereign power: 'The *coup d'état* is an assertion of *raison d'état*, and a self manifestation of the state' (Foucault 2007: 262). In the combat myths from out of which monotheism arose, however, Lucifer's rebellion is the most disturbing because, according to Christianity's defining account of that myth, it is made in the name of (prideful) equality not of (truer or more effective) sovereignty (Forsyth 1987, 2003). It is the claim to equality founded in the singularities of a heterogeneous world that more fundamentally threatens the sovereign machine.

The infinite government of finite things

To reign is to govern. To do either or both is to secure. If, however, reigning (that is to say sovereignty) is to be distinguished from government; if, according to Foucault, 'government finds its ends in the "things" to be directed' (Foucault 2007: 87), then the principle of formation of rule, together with the government of things and ends to be directed, are posed differently by the spatio-temporal conditions of possibility and operability set by the theologico-political and the modern, factically finitudinal, problematisation of politics. What perhaps most distinguishes the factically finite problematisation of politics, government and rule from its soteriological counterpart is the infinity of finite rule that it sets for modern politics. It is not simply this end, or that end, which has to be directed in and by factically finite politics, government and rule. If government finds its ends in the thing to be directed – and since nothing can be left to God here, except for modern capitalists the God of 'capital' – it is factical finitude *tout court* that presents itself to modern politics, government and rule.

Consequently, if factical finitude is the indefinite coming and going of an infinity of finite things, then rule becomes not simply the indefinite government of finite things, but the infinite regulation, also, of their very processes of coming and going. Manifestly, however much the factical world is given, even in its historicity, it is not given in any final shape or form. It is not only a dynamic flux. We shall see that understandings of what it is to be a living thing have undergone profound transformation and change during the course of the digital and molecular revolutions of the twentieth and twenty-first centuries. Where living entities are concerned it is not so much the fixed properties of the infinity of things that come to presence in existence, so much as the dynamic processes of interactive becoming that they exhibit. It is these which now attract the attention of the factically finite compulsion towards the infinite government of finite things.

While this book largely concerns the politics of Life, as life is construed biopolitically in the modern age, it does not ignore the intimate association of biopower with sovereign power. Equally, also, while it is concerned with what Foucault described as the urge to 'make life live', it also recognises with him that there is no life without death, and that what is addressed biopolitically is a certain construal of the life–death nexus for the purposes of determining novel rules of truth and truths of rule. Contrary to many early accounts, such as that offered by Hobbes – of how the problematisation of politics, government and rule, posed in factically finite terms, offered an escape from endless religious conflict and civil war – neither life nor death is a datum.[8] What they consist in depends upon many complex and historically changing practices of relation and differentiation. Hence, just as life is a kind of death for the Christian, so also death can never be dissociated from life, or indeed overcome in life, for the factically finite. That is the reason why the championing of natality over mortality, to which thinkers from Arendt to Esposito sometimes appeal, cannot ultimately work (Arendt 1958; Esposito 2008; O'Byrne 2010). Birth is no more a datum than life or death. There is natality only because there is mortality. Factically speaking, birth, life and death are indissociable. They differentiate and define one another in scientific as much as symbolic and semantic ways. To elevate natality over mortality ultimately refuses the challenge of factical freedom as much as does elevating mortality over natality. There is no natal fix, then, to the bind of factical freedom unto death. The birth of the new entails the death of the old.

Suffice to say for the moment, also, that the political problematisations of factical finitude did not arrive fully formed. Nor have they reached some historically determined outcome. Neither in fact, of course, have those of the theologico-political. The horizon of possibility and operability set by factical finitude is a manifold not a monolith. So also is that of the theologico-political. Factical finitude is a site of contestation to and for itself. So also is that of the theologico-political. Revealed religion has not solved the theologico-political problematisation of politics, government and rule that it has posed for millennia. The 'word' (*logos*) is not the last word, otherwise why would it have resulted in such a monumental word-making machine? Neither has factical finitude solved the factical problematisation of politics, government and rule that it has posed.

In sum, then, there is human finitude cast within the enframing of salvation – soteriological finitude. On the other hand, there is human finitude cast within the mere facticity of human existence – factical finitude. The exigencies of factical finitude open an entirely different problem space for the determination of human conduct, specifically its rules of political truth as well as its truths of political rule, than that offered by soteriological finitude. It is the argument of this book that biopolitics combined with modern sovereignty were two of the principal ways in which concrete political form has been given to human finitude, as this became cast in factical rather than soteriological terms. And that the prevailing character of the style, affects and effects of that move are predominantly, also, baroque. Throughout what follows I assume an intimate relation between modern sovereign and biopolitics. The historical evidence presents them emerging together out of the welter of change that characterised the baroque character of seventeenth-century Europe, its world making and its colonising. There are, however, additional, analytically compelling reasons for keeping sovereign and biopower in view throughout what follows.

Sovereignty and government

> If every unity implies a duality and therefore the possibility of uproar, or stasis, is immanent, then theology seems to become 'stasiology'.
>
> (Schmitt 2008b: 126)

Consider this opening argument, then, that modern sovereignty and biopower were two of the principal means by which concrete political form was given to modern factical finitude. Confronted with what Foucault called the crisis in the 'conduct of conduct' that characterised Europe and its colonies during the course of the sixteenth and seventeenth centuries, factical finitude progressively became the space of problematisation in which questions of politics, government and rule were posed and answered. As Foucault also observed, while modern sovereignty dominated the early modern political imagination, other kinds of power relations were simultaneously also instituted, practised and refined: reformed versions of pastoral power; police power, or *polizeiwissenschaft*; population management, which I would date earlier than Foucault and note its occurrence in European colonial as well as metropolitan rule; and so on.

Schmitt's political theology provides us with the best available contemporary theorisation of the modern sovereign conception of power, and its genealogical dependence upon, the theological absolutism of Christianity's *via moderna*. That theorisation exposes how sovereignty – comprised of an originary and ineliminable fracture – institutes the very idea of politics, government and rule as a political stasiology. Whereas the stasiology to which Schmitt refers arises in respect of his argument with Peterson's invocation of Christology and the Trinity, as that which defeats the very possibility of political theology (Schmitt 2008b; Peterson 2011), the fracture to which I am referring is different, but it nonetheless haunts Schmitt's texts, notably *The Concept of the Political* (2007), *Political Theology* (2005) and *Political Theology II* (2008b).

This fracture lies within the component parts of the sovereign conception of power itself. Only by dividing time into the temporality of the instant (of the miracle-like event of decision) and the temporality of succession (of the unproblematic legislation of sovereign will into the historical actuality of a juridico-political order) can the sovereign myth that sovereigns reign and governments rule be posited and sustained. This artifice elides every problematic binary distinction that it institutes, and with which historical orders of sovereign power have struggled since time immemorial: metaphysics versus mechanics; hierarchy versus heterarchy; reigning versus ruling; sovereignty versus government; strategy versus tactics; policy versus administration; friend versus enemy; inequality versus equality.

The fracture is a compound one that runs, first, between the metaphysical binaries of: metaphysics and mechanics; essence and existence; contingency and necessity. These fractures cannot arise sequentially, as if it were possible to have sovereignty and then government, or to have metaphysics and then mechanics, or essence and then existence. Technicity, Stiegler reminds us, is originary (Stiegler 1998, 2008). There can be no essence without its existence; no principle without its process; no emanation without its liturgical acclamation; no spectacle without its impresarios and stage managers. No reigning, in short, without ruling; no sovereignty without government. Sovereignty also poses these fractures as things to be violently overcome so that the relationship between rulers and the ruled may conform to one designed to subordinate and recruit the ruled into the endless war of normalising exception required for the continuous reinstitution of the sovereign. Sold as a protection racket, sovereign securitising always aims at securing sovereignty. Since sovereignty cannot in fact be secured its failure refuels the drive to sovereign power.

It runs secondly, however, between rulers and ruled. But I am less concerned with that equally compound fracture here. I wish to draw attention for the moment, instead, to the critical fracture that arises within the very order of the sovereign itself. For if the sovereign is not a homogeneous unity, if the sovereign is instead an unstable heterogeneity comprised of elements that do not add up, then an entirely different problematisation of politics, government and rule arises than that offered by the stasiology of modern theologico-political sovereignty. For that stasiology is itself, by definition, radically violent and unstable.

To drive home the point, consider also Rheinhart Koselleck's summary of the early modern institution of sovereignty. It supplements Schmitt by recalling that it was no mere juridical abstraction but, to use a favourite Schmittean expression, a concrete historical formation: 'The absolute ruler recognised no authority over himself than God, whose attributes in the political and historical field he appropriated' (Koselleck 2000 [1988]: 17). Other operational entailments followed logically from the conceptualisation and institutionalisation of modern sovereignty, and these compounded the inability of Schmitt's fractured sovereign to act the sovereign in the way demanded of it. 'To meet his all-encompassing responsibility,' observes Koselleck, 'the prince had to seek the measure of his actions in their calculable effect on everyone else. The compulsion to act thus provoked a need for heightened foresight. A rational calculation of all possible consequences came to be the first political commandment' (Koselleck 2000 [1988]: 20).

Strictly speaking, Schmitt's sovereign seems to be no strategist. His *metier* is the 'miracle' of decision. And yet his conditions of possibility set conditions of operability that simultaneously also oblige his will to be strategically informed as well. However it may be said to operate philosophically, there is no escaping the originary, mutually disclosing, link between political *phusis* and *technics* (Stiegler 1998, 2008). Specification of the state of emergency, differentiation of friend from enemy and preparedness to institute public killing to make it so, all appear to be composite aspects of a single – a singular – political will. But, then, how is the friend/enemy distinction to be drawn? How is the enemy to be securely identified? Will any friend suffice? Will any enemy do? What if you misconceive the friend? What if your friends deceive you? What if you misconceive the enemy? What if there is more than one existential alien with which to contend? Forget the difficulties of positivistically posing let alone answering such questions. What if the relationship between friend and enemy, and the miracle of decision for which it calls, is radically undecidable (Derrida 2005)? But let's leave the Derridean challenge and stay with the one posed to the miracle of decision by its inescapable and positivistic need to know.

To be sovereign, to become sovereign, requires the exercise of a strategic calculus of necessary killing to resolve these and other related issues, not least those to do with timing and interpreting, what Machiavelli called, the signs of the times. One that cannot be elided, in the least, is the following additional question: How much killing is enough to satisfy the strategic calculus for which the sovereign differentiation of friend from enemy and the institution of a stable juridico-political order calls?

The sovereign's strategic calculus of necessary killing does not arise before the decision is made. It cannot arise after the decision is made. It has, complexifying the 'miracle' of decision, to be incorporated within the decision as such. How much killing is enough may or may not be explicitly included as part of the strategic calculus of necessary killing. It has nonetheless to be answered somehow in the moment of decision, since it is simultaneously posed by that decision, if the regime of the modern sovereign is not to become that of a mere abattoir. (You can be forgiven for thinking that, in general, it is.)

These are precisely the problems to which Koselleck refers. And they were recognised from the very beginning of the institution of the concepts and practices of modern sovereigns. They were, for example, the very problems explored in German *Trauerspiel* of the late seventeenth century (Benjamin 2009 [1925]). Charged not only with exercising a strategic calculus of necessary killing, but also with answering the question of how much killing is enough, every human sovereign is condemned to fail the test of sovereignty itself. Forget the rational pursuit of strategic ends. Forget, for example, the strategists' recent condemnation of Bush and Blair for their irresponsible sovereign decision-making over Iraq. Strategists do not have the answer either. For no such strategic calculus of necessary killing is humanly available, and none is thereby capable of specifying how much killing is enough to realise the sovereign ends that such a calculus presupposes. When asked, how much killing is enough?, the sovereign always answers more (cf. Tony Blair's call for military intervention against 'Islamic extremism' in general and into the Syrian civil war in particular: *The Guardian*, 13 June 2013; *The Daily Telegraph*, 8 July 2013, *Mail Online*, 15 June 2014).

Again, Koselleck references, historically, a point that is a logical entailment of the very conceptualisation of political power as sovereign. The modern sovereign, he says,

> accumulated all responsibility … but to maintain it he had to remain aware of the augmented responsibility it entailed … Then, and only then, was he in possession of the authority that guaranteed his power … [and] came to feel a compulsion to act which continually conjured up new decisions … the consequences of inaction could be as serious as those of its opposite. One risk matched and constantly challenged the other. Indeed the danger of falling from one extreme into the other was the very source of the evidence of sovereign decisions.
>
> (Koselleck 2000 [1988]: 20)

However, he concludes: 'in order to keep the consequences of his actions (which once committed were not humanly alterable) under his own control as long as possible, the prince was driven to augment his power, which in turn increased the sources of danger, the risk of abusing or failing to use the accumulated power' (Koselleck 2000 [1988]: 20). The miracle of decision is therefore also comprised of 'a logic of absolute responsibility' (Koselleck 2000 [1988]: 20) the baroque discharge of which can only ever be humanly simulated rather than exercised.

Reigning and ruling therefore comprise no sequence. Theirs is also a reversible relationship. *Coup d'état* reminds us of the very point (Carlton 1997; Foucault 2007). To reign is to rule and to rule is to reign. However conceptually heterogeneous they may be, and they are, reigning and ruling are locked in a chiasmic relationship. They are not only reversible – Praetorians, Janissaries, Mamluks, military chiefs and cameralist administrators having done so for millennia – they are also subject to the mediation of a third term. That third term is 'politics'. A politics un-thought by Schmitt, because it is the un-thought of the very rules of truth and truths of rule of theological absolutism from which the modern account of sovereignty derives.

This excursus into the originary political stasiology of the sovereign conception of power is, therefore, designed to reinforce the following point. The modern sovereignty of which Schmitt speaks is a fractured compound comprised of heterogeneous terms. It is a compound which is required to institute burgeoning states of emergency, whose institution requires a degree of strategic calculation that exceeds its mortal powers. Sovereigns fail. They fail serially. And when exercising the very power over death, of which they are uniquely comprised, they fail utterly. They are haunted by the continuous threat of such failure. It accounts for the baroque pathos, and the bathos, that surrounds the attempted exercise of sovereign power.

Schmitt's sovereign is constitutively incapable of doing what it is supposed to do, and remaining what it is supposed to be. Here the sovereign's rule of truth and truth of rule must ultimately be rendered also as the positivistic knowing (*connaissance*) of modern strategic thought. The burgeoning of strategic thinking, and its will to positivistic knowledge promises an escape from this aporia, but singularly fails, of course, to deliver the sovereign from it. It becomes an integral part of that great baroque spectacle of simulation that the apparatus of modern sovereignty demands in order to feign the fiction of sovereign power. The appetite for strategic *connaissance* – how else is the existential enemy to be objectified, its properties positivistically mapped, and its threat exposed so that it can be infinitely overcome? – is endlessly stimulated by the appetite for miraculous decision-making. The one stimulates and services the other in resonances as sexual and aesthetic as they are political. Inevitably, each condemns the other for failing to deliver.

Why then was sovereignty alone not sufficient to answer the questions of politics, government and rule posed by the move to factical finitude, offering relief from the threat of violent death at the hands of other men while becoming an engine for the mobilisation of mass killing? The answer offered here is that sovereignty is a metaphysics of rule that is always already wanting in mechanics. It offers an onto-theological principle of rule not an apparatus of government. But it cannot do without one. All sovereigns require the means of operationalising their sovereignty. To will death politically – it is what the sovereign consists in – is to open up a political economy of killing. But merely to open up a political economy of killing is always already to call for its enactment. Schmitt was correct: sovereignty is a killing machine, but its baroque instrumentalisation of political death is not only fated to fail, the sovereign itself is always exposed and transformed in the act of that killing.

As Foucault noted, government does not secure its means and ends from sovereign will alone. It finds its means and ends according to the properties of that which it seeks to govern. This sometimes blows back to govern sovereigns as much as sovereigns seek to govern through government (recall the threat of *coup d'état*).[9] Government does not, of course, refer simply to 'the government', the cabinet, chamber or *camera* comprising courtiers or ministers appointed to do the sovereign's will, together with the intrigue and machination that characterises them. It comprises all of the mechanisms by which the conduct of conduct is regulated and self-regulated. Biopolitics provides some of the mechanics that sovereignty requires. It too endangers the myth of sovereign rule, however, more so today than ever before since it threatens to become everything that rule consists in. Schmitt's fear was precisely this. And he was right about it, if he was nonetheless wrong about the presumed efficacies as well as the political virtue of sovereign necessity.

Pascal without the wager

> Cast into the infinite immensity of spaces, of which I am ignorant, and which know me not, I am frightened.
>
> (Blaise Pascal)

It was, therefore, not the early modern political rationality of *raison d'état*, alone, that posited that henceforth man would have to live in an indefinite time. Since the beginning of modernisation, men and women subject to it have been living in such a Pascalian universe (Campe 2012; Hacking 1975, 1990; Daston 1988). As the challenge to give concrete form to factical rather than soteriological finitude emerged, Foucault among many others recognised that it did so in the context of a political temporality, 'with specific characteristics in comparison with the temporality that dominated the thought of the Middle Ages, and even of the Renaissance, because it is an indefinite temporality' (Foucault 2007: 255). Such a temporality came to pose the problematisation of politics, government and rule in specifically different ways. Indeed it was embraced by states, and emancipatory political movements alike, as a means of freeing themselves from the scholastic mentalities as well as papal, royal and feudal institutions and practices of government and rule. These also included, of course, the brute domination and extractive accumulation that gave concrete form to the preceding, soteriological, problematisation of government and rule. It is this novel problematisation of politics, government and rule that constitutes the political analytic of a finitude in which an entirely new challenge is issued to the problematisation of the nature of nature *tout court*, as much as politics and knowing. Since, with the retreat of God, Nature became deficient, 'there is *lack* in Nature' – in all natures – and, '*because of that very fact*, something is *added* to it' (Derrida 1997: 244, emphasis in the original). Herewith, then, the necessity of the artifice and personating that Hobbes, too, found so essential to that early modern state building which initially found its expression through the baroque (Maravall 1986). Henceforth, the political task of factical finitude becomes that of specifying the proliferating truths, rationalities, arts and technologies required for the indefinite government and rule of an infinity of finite things, and their processes of becoming. In short, it becomes that of securing concrete political form and specular expression for factical finitude.

Thus, while the analytic of finitude progressively ceased to proclaim the coming of the last day, a revised eschatology and a revised katechontic task also came to define the task of modern government and rule. For in the infinity of finite things that characterises the political analytic of modern finitude, the one thing of which you can be sure is that nothing lasts. Everything finite comes to an end. No last day, then, but every day is potentially an end of days for the finite, not least the politically finite. Thus the eschaton that once differentiated between the finite and the transcendentally eternal is now continuously imminent as well as immanentised within the infinity of finite things. Subsequently, the political subject becomes the political spectator, whose infinite servility and voluntary servitude to indefinite government is assiduously cultivated by democratic and dictatorial regime alike in the name of their security.

Herein, also, and paradoxically, a new, relentless because indefinite, katechontic imperative is levied upon modern government and rule. Warranted by factical finitude, finite regimes of government and rule have indefinitely to secure themselves against the realisation of their very own finitude. And it is this infinite task, differently pursued in sovereign and biopolitical *dispositifs de sécurité*, that henceforth founds the baroque modernisation of politics, government and rule of Man, of the State and now also of Life, within the temporal horizons and imperatives set by the modern.

As Foucault also observed of the immanent eschatology of this Pascalian cosmos: 'There is something here profoundly bound up with our Modernity: apart from its religious moralities' (Foucault 1989: 357). Something about the very analytic of finitude, comprised also of an infinite oscillation between the Same and the Other, institutes an immanently eschatological rupture, katechontically policed, that now defines its biopolitical as well as sovereign security politics. Indeed that unites them in a web of infinite securitisation of finite things. Scipio's dream thus haunts every regime of modern government and rule. Their finitude is their very warrant to rule, their very finite artifices the means of compensating for Nature's lack. It is, however, precisely, also, that finitude which ordains that there will be a time when they will have ended. Their warrant to rule nonetheless equally, therefore, ordains that they must rule indefinitely. None seek the end of their government and rule, only the indefinite pursuit of the infinity of finite ends, and the infinite securitisation of finite things, through which government and rule has to be operationalised and legitimated. Indeed contemporary biopolitics of security now targets the infinity of processes of emergence as much as the infinity of the finite things that come to presence through them (Dillon and Reid 2009).

In as much as the analytic of finitude problematises government and rule as the infinite pursuit of the regulation and administration of finite processes and things, modern regimes of government and rule become terrorised by the immanence and imminence of their own demise. After all, government and rule are themselves not only finite. Their very operational conditions of government and rule, as well as the warrants they take to govern and rule, are also instituted by modern factical finitude. Terrorised by the immanence and imminence of their own demise, their politics of security katechontically assert an opposite but equally absolute right to repress, suppress and terrorise in their turn anything that threatens to realise their finitude. They also violently insist on an indefinite duty to obey the dictates of these politics of security *in extremis* – and, of course, they are by definition always already, factically, *in extremis* (M. Dillon 2007)

'Man', as Foucault taught, was thus the strategic figure initially invoked to stabilise the epistemic field of formation giving positivity to the infinity of finite things of which modern factical finitude is comprised. Life, I also note, soon emerged as another empirico-transcendental double to rival and later dominate Man, giving concrete political and governmental form to factical finitude in different ways, not least through prioritising the by now ubiquitous, diversely interpreted, multiform and classically baroque trope of 'the event' or 'marvel' (Dillon 2013).

Since the properties of the figure of Man were so evidently drawn from the figure of God, Man is God's after-life. Since sovereignty (Man) and biopolitics (Life) were intimately allied from their very modern inception, the Life of modern biopolitics arose as a means of compensating for the inability of Man to satisfy the many governmental demands for the modernisation of politics, government and rule in the after-life of the death of God, and the advent of modern factical finitude. When Man, stillborn of God, entered the scene of modern politics, he had already therefore to be buttressed by governmental concern with modern accounts of Life, Life in the form of the emerging properties of species being, of which population was one early positivistic referent. It was against this living death that Foucault went on to interrogate the trope of spirituality, and specifically also that of the political spirituality manifested in the political madness of the insurrection in Tehran in 1978 – 'a strike *in relation to* [modern] *politics*' (Foucault, in Afary and Anderson 2005: 212; Dillon 2013).

Henceforth, therefore, living the analytic of finitude in terms of government and rule, the positivisation of the infinity of finite things became an indefinite task that progressively took not only population and species being, for example, but also the contingency, probability and eventalness, with which life as species being was also correlated, as definitive referent objects for the operationalisation of modern government and rule as much as knowing. Within modernity's political analytic of finitude, the very accident of Life, as such, especially, was so radicalised and transformed by the mathematisation of risked-based calculations to infinity – as much as by the allied understanding of complex connectivity, the irredeemable opacity and retreat of origins, and the ubiquity of unintended outcomes – that the modern sciences of the contingent have become a central feature of the biopolitical governmentalisation of the late modern state as much as the '*étatization*' of late modern government (Hacking 1975, 1990; Daston 1988; Gigerenzer *et al.* 1990; Baker and Simon 2002). In the process the very rule of truth and truths of rule associated with *tuche* and chance, then *fortuna* and subsequently thereafter with probability, risk and contingency were transformed in the process.

According to Foucault, under the Classical episteme, '[t]he order of nature was conceived, prior to any catastrophe' (Foucault 1989: 358). According to the biopolitical political analytic of finitude, the advent of Life becomes, instead, the product of a cosmic catastrophe. Living things become the indefinite emergency of the complex processes of emergence precipitated by that catastrophe. The construal of Life, positivistically in molecular and digital terms, as the infinity of finite becomings, institutes *dispositifs de sécurité* preoccupied with the indefinite biopolitical government of emergencies of emergence. These newly formed neo-baroque arts of biopolitical governance are defined in terms, for example, of 'chaos' as well as 'resilience' (Zebrowski 2014).

Foucault noted that modern government and rule knows no final end. It is governed by an infinity of finite ends … and finite endings. Everything comes and goes including infinity, since, after Cantor, modern maths teaches us that there is an infinity of infinities (Campe 2012; Clegg 2003). Factical finitude is then the fate of modern government and rule, as much as it is its field of problematisation. What other features therefore characterise its response to that fate?

Given the nature of its tasks – infinite securing of finite processes and things – modern government and rule is, of course, bound to fail. For who knows how to resolve its problem? Its responses nevertheless extend beyond that of the endless remediation of failure. In his epistemic analytic of finitude, Foucault offers several clues to how they do so.

For example, continually seeking to push back the frontiers of knowing, modern knowing does so only to encounter the unknowable other of its knowing. Its epistemic objects retreat before its positivistic gaze, endlessly proliferating as they simultaneously also recede into an alterity whose very opacity excites its epistemic adventures (Rheinberger 1997, 2010; Knorr-Cetina 1999). The positivistic knowing that radically empowers and refuels its awesome command of natural and social worlds, and their intersection, retreats before the very power of its gaze. The more it knows, the less it knows, and the more it is impelled by the will to know. Modern thought, Foucault explains,

> both for itself and in the density of its workings, should be both knowledge and a modification of what it knows, reflection and a transformation of the mode of being of that on which it reflects. Whatever it touches it immediately causes to move: it cannot discover the un-thought, or at least move towards it, without immediately bringing the un-thought nearer to itself.
>
> (Foucault 1989: 357)

Nearer to, but never comfortably or coherently within, itself. Consider, only, the example of modern data gathering by the US National Security Agency and its allied intelligence agencies such as Britain's GCHQ.

Modernity's political truth – the infinite securing of finite government and rule, the endless task of rendering materiality governable – similarly also retreats before modernity's political gaze. The infinite intractability of the materiality to be governed, in its constantly discoverable new processes and forms, and its correlate resistance also to rule, even as it proves infinitely subject and amenable to rule, is precisely that with which modern government and rule must endlessly contend. Just as counting proves the existence of infinity so also the one number that can never be known is that of the infinite (Campe 2012; Clegg 2003). Everything is therefore dangerous because, in its very know-ability, which is to say its governability, everything also proliferates ungovernably into the infinity of its factically emergent finite materiality. This modern rendering of materiality resists, therefore, less out of obdurate will, though there is always of course that, than out of the very indefinite infinity of its finitudinal facticity.

The more radically transparent modern rulers and the ruled become to demographic and digital knowledge, the more politically opaque does the world become to rulers and the ruled alike. Since it is the governors who most intimately understand the fallibility and fragility of pursuing the infinite securing of finite things – the ruled suffer it – the opacity of modern political truth, and the serial life of governmental failure, is equally well known professionally to them. They encounter and experience it endlessly. Hence, the baroque pathos and bathos, along with manufactured, mannered and mediatised enthusiasms, that continually inflect their tone. In its pursuit of the infinite securing of finite things, one of the most distinguishing features of modern government and rule is, therefore, less correction than *ressentiment*. It is a *ressentiment* expressed by those who govern, or those whose sensibilities have been entirely colonised by a prevailing *mentalité du gouvernement*, towards the infinity of finite things that always already exceed government and rule, even as they are endlessly presented to be governed. Among other things, that *ressentiment* fuels their relentless cruelty and violence.

Notes

1 Jonathan Swift, 'On Poetry', in Swift (1958).

2 Infinity: ∞ 'It is my task to convince you not to turn away because you don't understand it. You see, my physics students don't understand it either. That is because I don't understand it. Nobody does.' Physicist Richard Feynman quoted in Clegg (2003: 3). Ellipsis: … 'Mathematicians, more fussy than the rest of us, take it specifically to mean, "and so on without any end"' (Clegg 2003: 8).

3 A modern rather than a classical account of living being. As Charles Shepherdson observed in relation to, what we call, the Epic as well as the Tragic age of the Greeks: 'The body (*soma*) in epic is neither "extended substance" nor the site of appetite and desire that the viruous man will regulate by his reason, but the physical remainder that lies on the earth once the warrior's spirit has left it .... And obviously, the concept of "soul" (*psuche*) whose history one can trace across epic, tragedy, and philosophy (and with it the history of subjectivity) from *psuche* to *animus* to *spiritus*, and from Latin formations to the Christian "soul", to Teutonic "*Geist*" and the Anglo-Saxon "mind", and then to the "brain" and "consciousness and the Freudian "ego" ... will shift alongside these changing conceptions of the body"' (2009: 51). The same may be said of the careers of *bios* and *zoe* (Bultmann *et al.*).

4 I am adapting what Erich Auerbach had to say about cultural interpretation to politics. See Kahn (2014: 8).

5 For a useful survey of the literature on, and reappraisal of the controversial figure of, Eusebius, see Hollerich (1999); Cunningham *et al.* (2008).

6 Derrida (1994, 2005) offers an entirely different account of decision.

7 Technicity (mechanics here) is originary to metaphysics. See, for example, Stiegler's argument with Heidegger and Derrida (1998, 2008).

8 Death in tragedy is, for example, different from death in epic, as Nicole Loraux has shown in *Tragic Ways of Killing a Woman* (1987). The way in which life and death are conceived now cannot simply be inserted into the broader, and radically different, environment of 'Greek culture', 'Greek thought', 'Greek law' or 'Greek politics' without considerable violence to the Greeks, as much as to ourselves (Shepherdson 2009).

9 Foucault is careful to note that *coup d'état* is part of the sovereign game. The idea is to install a more effective sovereign. Jacques Rancière brilliantly analyses how the very order of accounting that sovereign rulers seek to install is contested politically (Rancière 1999).

Bibliography

Adler, Emanuel (1998) 'Condition(s) of Peace', *Review of International Studies*, 24(5): 165–91.

Afary, Janet and Anderson, Kevin (2005) *Foucault and the Iranian Revolution: Gender and the Seductions of Islamism* (Chicago, IL: University of Chicago Press).

Agamben, Giorgio (2011) *The Kingdom and the Glory: For a Theological Genealogy of Economy and Government* (*Homo Sacer* II, 2) (Stanford, CA: Stanford University Press).

Agamben, Giorgio (2013) *The Highest Poverty: Monastic Rules and Form-of-Life* (Stanford, CA: Stanford University Press).

Arendt, Hannah (1958) *The Human Condition* (Cambridge: Cambridge University Press).

Auffarth, Christoph and Stuckenbruck, Loren T. (2004) *The Fall of the Angels* (Leiden: Brill).

Baker, Tom and Simon, Jonathan (2002) *Embracing Risk: The Changing Culture of Insurance and Responsibility* (Chicago: University of Chicago Press).

Benjamin, Walter (2009 [1925]) *The Origin of German Tragic Drama* (London: Verso).

Bernauer, James (1999) 'Foreword', in Michel Foucault, *Religion and Culture*, ed. Jeremy R. Carette (London and New York: Routledge), pp. xi–xviii.

Bultmann, Rudolph., Von Rad, G. and Bertram, G (1965) *Life and Death* (London: Adam and Charles Black)

Campe, Rüdiger (2012) *The Game of Probability: Literature and Calculation from Pascal to Kleist* (Stanford, CA: Stanford University Press).

Carlton, Eric (1997) *The State against the State: Theory and Practice of the Coup d'Etat* (London: Ashgate).

Cavanaugh, William T. (2002*) Theopolitical Imagination* (London and New York: T. & T. Clark).

Caygill, Howard (2013) *On Resistance: A Philosophy of Defiance* (London and New York: Bloomsbury Academic).

Chamayou, Grégoire (2012) *Manhunts: A Philosophical History* (Princeton, NJ: Princeton University Press).

Clark, Timothy (2005) *The Poetics of Singularity: The Counter-Culturalist Turn in Heidegger, Derrida, Blanchot and the Later Gadamer* (Edinburgh: Edinburgh University Press).

Clegg, Brian (2003) *A Brief History of Infinity: The Quest to Think the Unthinkable* (London: Robinson).

Cunningham, David S., Cory, Catherine and Hollerich, Michael (2008) *The Christian Theological Tradition* (London: Prentice Hall).

Daston, Lorraine (1988) *Classical Probability in the Enlightenment* (Princeton, NJ: Princeton University Press).

Debord, Guy (1994) *The Society of the Spectacle* (New York: Zone Books).

Delumeau, Jean (1990) *Sin and Fear: The Emergence of a Western Guilt Culture, 13th–18th Centuries* (New York: St. Martin's Press).

Derrida, Jacques (1973) *Speech and Phenomena, and Other Essays on Husserl's Theory of Signs* (Bloomington: Indiana University Press).

Derrida, Jacques (1994) *Specters of Marx: The State of the Debt, the Work of Mourning, and the New International* (London and New York: Routledge).

Derrida, Jacques (1997) *Of Grammatology* (Baltimore: Johns Hopkins University Press).

Derrida, Jacques (2005) *The Politics of Friendship*, trans. George Collins (London: Verso).

Dillon, Michael (2007) 'Governing Terror: The State of Emergency of Biopolitical Emergence', *International Political Sociology*, 1(1): 7–28.

Dillon, Michael (2011) 'Specters of Biopolitics: Finitude, Eschaton, and Katechon', *South Atlantic Quarterly*, 110(3): 780–92.

Dillon, Michael (2013) 'Afterlife: Living Death to Political Spirituality', *Millennium – Journal of International Studies*, 42: 114–34.

Dillon, Michael and Reid, Julian (2009) *The Liberal Way of War: Killing to Make Life* (Abingdon: Routledge).

Dillon, Sarah (2007) *The Palimpsest: Literature, Criticism, Theory* (London and New York: Continuum).

Dupré, Louis K. (1993) *Passage to Modernity: An Essay in the Hermeneutics of Nature and Culture* (New Haven: Yale University Press).

Egginton, William (2010) *The Theater of Truth: The Ideology of (Neo)Baroque Aesthetics* (Stanford, CA: Stanford University Press).

Esposito, Roberto (2008) *Bios: Biopolitics and Philosophy* (Minneapolis: University of Minnesota Press).

Evans, Brad and Reid, Julian (2014) *Resilient Life: The Art of Living Dangerously* (Cambridge: Polity Press).

Forsyth, Neil (1987) *The Old Enemy: Satan and the Combat Myth* (Princeton, NJ: Princeton University Press).

Forsyth, Neil (2003) *The Satanic Epic* (Princeton, NJ: Princeton University Press).

Foucault, Michel (1989) *The Order of Things: An Archaeology of the Human Sciences* (London and New York: Routledge).

Foucault, Michel (2001) *Fearless Speech*, ed. Joseph Pearson (New York: Semiotext(e)).

Foucault, Michel (2003) *Society Must Be Defended: Lectures at the Collège de France, 1975–76* (London: Picador).

Foucault, Michel (2006) *History of Madness* (London and New York: Routledge).

Foucault, Michel (2007) *Security, Territory, Population: Lectures at the Collège de France, 1977–78* (Basingstoke: Palgrave Macmillan/République Française).

Foucault, Michel (2010) *The Government of Self and Others* (Basingstoke: Palgrave Macmillan).

Foucault, Michel (2011) *The Courage of Truth: The Government of Self and Others II: Lectures at the Collège de France, 1983–84* (Basingstoke: Palgrave Macmillan).

Genette, Gérard (1997) *Palimpsests: Literature in the Second Degree* (Lincoln: University of Nebraska Press).

Gigerenzer, Gerd, Swijtink, Zeno, Porter, Theodore, Daston, Lorraine, Beatty, John and Krüger, Lorenz (1990) *The Empire of Chance: How Probability Changed Science and Everyday Life* (Cambridge: Cambridge University Press).

Gillespie, Michael Allen (2008) *The Theological Origins of Modernity* (Chicago, IL: University of Chicago Press).

Hacking, Ian (1975) *The Emergence of Probability: A Philosophical Study of Early Ideas about Probability, Induction and Statistical Inference* (Cambridge: Cambridge University Press).

Hacking, Ian (1990) *The Taming of Chance* (Cambridge: Cambridge University Press).

Harding, Luke (2014) *The Snowden Files: The Inside Story of the World's Most Wanted Man* (London: Faber and Faber).

Harries, Karsten (2001) *Infinity and Perspective* (Cambridge, MA: MIT Press).

Hollerich, Michael J. (1999) *Eusebius of Caesarea's Commentary on Isaiah: Christian Exegesis in the Age of Constantine* (Oxford: Oxford University Press).

Honig, Bonnie (2008) 'Review Essay: What Foucault Saw at the Revolution on the Use and Abuse of Theology for Politics', *Political Theory*, 36: 301–12.

Jonas, Hans (1952) 'Gnosticism and Modern Nihilism', *Social Research*, 19(1): 430–52.

Jonas, Hans (2001) *The Gnostic Religion: The Message of the Alien God & the Beginnings of Christianity* (Boston: Beacon Press).

Kahn, Victoria (2014) *The Future of Illusion: Political Theology and Early Modern Texts* (Chicago, IL: Chicago University Press).

Kantorowicz, Ernst Hartwig (1946) *Laudes Regiae: A Study in Liturgical Acclamations and Mediaeval Ruler Worship* (Berkeley, CA: University of California Press).

Kantorowicz, Ernst Hartwig (1998) *The King's Two Bodies: A Study in Mediaeval Political Theology* (Princeton, NJ: Princeton University Press).
Kaplan, Robert and Kaplan, Ellen (2003) *The Art of the Infinite: The Pleasures of Mathematics* (Oxford: Oxford University Press).
Kauffman, Stuart A. (1993) *The Origins of Order: Self-Organization and Selection in Evolution* (Oxford: Oxford University Press).
Kauffman, Stuart A. (1995) *At Home in the Universe: The Search for Laws of Self-Organization and Complexity* (Oxford: Oxford University Press).
Kaufman, Peter Iver (1990) *Redeeming Politics* (Princeton, NJ: Princeton University Press).
Keck, David (1998) *Angels and Angelology in the Middle Ages* (Oxford: Oxford University Press).
Knorr-Cetina, Karin (1999) *Epistemic Cultures: How the Sciences Make Knowledge* (Cambridge, MA: Harvard University Press).
Koselleck, Reinhart (2000 [1988]) *Critique and Crisis: Enlightenment and the Pathogenesis of Modern Society* (Cambridge, MA: MIT Press).
Koyré, Alexandre (1957) *From the Closed World to the Infinite Universe* (Baltimore, MD: Johns Hopkins University Press).
Krüger, Lorenz, Daston, Lorraine, Heidelberger, Michael, Gigerenzer, Gerd and Morgan, Mary S. (eds) (1987) *The Probabilistic Revolution* (Cambridge, MA: MIT Press).
Lambert, Gregg (2008) *On the (New) Baroque* (Aurora, CO: Davies Group).
Lefort, Claude (1988) *Democracy and Political Theory* (Minneapolis: University of Minnesota Press).
Leigh, David, Harding, Luke, Pilkington, Edward, Booth, Robert and Arthur, Charles (2013) *WikiLeaks: Inside Julian Assange's War on Secrecy* (London: Guardian Books).
Loraux, Nicole (1987) *Tragic Ways of Killing a Woman* (Boston: Harvard University Press).
Maravall, José Antonio (1986) *Culture of the Baroque: Analysis of a Historical Structure*, trans. Terry Cochran (Minneapolis: University of Minnesota Press).
Metzger, Marcel (1997) *History of the Liturgy: The Major Stages* (Collegeville, MN: Liturgical Press).
Mezzadra, Sandro, Reid, Julian, and Samaddar, Ranabir (eds) (2013) *The Biopolitics of Development: Reading Michel Foucault in the Postcolonial Present* (New York: Springer).
Monod, Jacques (1997) *Chance and Necessity: Essays on the Natural Philosophy of Modern Biology* (Harmondsworth: Penguin).
Moore, R. I. (1987) *The Formation of a Persecuting Society: Power and Deviance in Western Europe, 950–1250* (Oxford: Blackwell).
Muir, Edward (1986) *Civic Ritual in Renaissance Venice* (Princeton, NJ: Princeton University Press).
Nagy, Gregory (1983) 'Sema and Noesis + Semiotics and the Classics: Some Illustrations', *Arethusa*, 16: 35–55.
Nancy, Jean-Luc (1991) *The Inoperative Community* (Minneapolis: University of Minnesota Press).
O'Byrne, Anne E. (2010) *Natality and Finitude* (Bloomington: Indiana University Press).
Pagden, Anthony (1993) *European Encounters with the New World* (New Haven, NT: Yale University Press).
Peterson, Erik (1964) *The Angels and the Liturgy* (New York: Herder and Herder).
Raffoul, François and Nelson, Eric Sean (eds) (2008) *Rethinking Facticity* (Albany, NY: State University of New York Press).
Raffoul, François and Pettigrew, David (eds) (2002) *Heidegger and Practical Philosophy* (Albany, NY: State University of New York Press).
Rancière, Jacques (1999) *Disagreement: Politics and Philosophy* (Minneapolis: University of Minnesota Press).

Reed, Annette Yoshiko (2005) *Fallen Angels and the History of Judaism and Christianity: The Reception of Enochic Literature* (Cambridge: Cambridge University Press).

Rheinberger, Hans-Jörg (1997) *Toward a History of Epistemic Things: Synthesizing Proteins in the Test Tube* (Stanford, CA: Stanford University Press).

Rheinberger, Hans-Jörg (2010) *On Historicizing Epistemology: An Essay* (Stanford, CA: Stanford University Press).

Rotman, Brian (1993a) *Ad infinitum … The Ghost in Turing's Machine: Taking God Out of Mathematics and Putting the Body Back In: An Essay in Corporeal Semiotics* (Stanford, CA: Stanford University Press).

Rotman, Brian (1993b) *Signifying Nothing: The Semiotics of Zero* (Stanford, CA: Stanford University Press).

Rotman, Brian (2000) *Mathematics as Sign: Writing, Imagining, Counting* (Stanford, CA: Stanford University Press).

Schmitt, Carl (2005) *Political Theology: Four Chapters on the Concept of Sovereignty* (Chicago, IL: University of Chicago Press).

Schmitt, Carl (2007) *The Concept of the Political* (Chicago, IL: University of Chicago Press).

Schmitt, Carl (2008a) *Constitutional Theory* (Durham, NC: Duke University Press).

Schmitt, Carl (2008b) *Political Theology II* (Cambridge: Polity Press).

Schwartz, Regina M. (2008) *Sacramental Poetics at the Dawn of Secularism: When God Left the World* (Stanford, CA: Stanford University Press).

Shepherdson, Charles (2009) 'Antigone: The Work of Literature and the History of Subjectivity', in Denise Eileen McCoskey and Emily Zakin, eds, *Bound by the City: Greek Tragedy, Sexual Difference and the Formation of the Greek Polis* (New York: State University of New York Press).

Stambaugh, Joan (1992) *The Finitude of Being* (Albany, NY: State University of New York Press).

Stiegler, Bernard (1998) *Technics and Time* (Stanford, CA: Stanford University Press).

Stiegler, Bernard (2008) *Technics and Time 2: Disorientation* (Stanford, CA: Stanford University Press).

Strayer, Joseph R. (2005) *On the Medieval Origins of the Modern State* (Princeton: Princeton University Press).

Swift, Jonathan (1958) *The Poems of Jonathan Swift*, 2 vols., ed. William Ernst Browning (London: Routledge & Kegan Paul).

Zebrowski, Chris (2014) *Value of Resilience: Securing Life in the 21st Century* (London and New York: Routledge).

2

BIOPOLITICS OF SECURITY IN THE TWENTY-FIRST CENTURY

> Freedom is nothing but the correlative development of apparatuses of security.
> (Foucault 2007: 48)

Introduction

This chapter addresses two questions. It first asks what happens to security practices when they take species life as their referent object. It then asks what happens to security practices which take species life as their referent object when the very understanding of species life undergoes transformation and change. In the process of addressing these two questions the chapter provides an exegesis of Michel Foucault's analytic of biopolitics as a collection of changing *dispositifs de sécurité* and contrasts this account of security with that given by traditional geopolitical security discourses. Biopolitics of security secures by instantiating a general economy of the contingent throughout all the processes of circulation which impinge upon the promotion of the re-productive powers and potentials of Life as species being. The chapter also theorises beyond Foucault when it interrogates the impact in the twentieth century of the compression of morbidity on populations and the molecular revolution on what we now understand life to be. It concludes that 'population' which was the empirical referent of early biopolitics is being superseded by 'heterogenesis'. This serves as the empirical referent for the recombinant biopolitics of security of the molecular age.

With that, and other transformations in the arts of artifice, the baroque becomes neo-baroque. Its very problematisation of the field of artifice, and not just its techniques, are substantially revised in the same way that modern biopolitics has been by the advent, for example, of digitalisation and molecularisation. Many of the established tropes remain - a world in crisis; the valorisation of the new; the world as stage; labarynthine political maneuvering; the world as diseased and disordered;

and so on. Such figures, terms of analysis as well as art, nonetheless found new expression. But others have entered the neo-baroque vocabulary of contemporary politics and culture as well. These include most significantly: apocalypse; emergency; eventalness; pluripotency; contingency; complexity; resilience; and heterogenesis. Hence 'the reality that the baroque was concerned about penetrating and dominating was not to be found in the Platonic realm of the essences but in the dramatic and changing sphere of life' (Maravall 1986: 257). In the baroque as in our neo-baroque biopolitics, 'the human does not possess a being that has become, but a being becoming - a *fieri*, not a *factum*, consequently a being incomplete and in continuous change' (Maravall 1986: 260).

By seeking to develop the theorisation of biopolitics with Foucault beyond Foucault, however, this chapter offers more than an exegesis of Foucault. It pursues this strategy by asking two questions.

The first, a collection of related questions, is initially posed by Foucault, when, in effect, he asked: What happens to the political rationalities and technologies of power, to the problematisation of security and to the character of security technologies when – taking species life as its referent object – power comes to strategise human being politically as species being? The chapter goes on to explain how, according to Foucault, biopolitics first evolved around 'population' as its empirical referent object of power. Given Foucault's emphasis on 'empiricities', any engagement of Foucault's biopolitics has therefore to address the category of the empirical referent of 'population', as well as the more generic referent object of biopower which is Life, and to ask what has happened to population as well as to Life in the interim, between Foucault's initial interrogation of the biopolitical economy of biopower and the biopolitics of the twenty-first century.

This chapter does precisely that.[1] It explains how the biopolitics of biopower is necessarily also allied with freedom, and what kind of 'freedom' is understood to be at work in it. In explaining precisely what Foucault understands, in addition, by 'security', and how this understanding of security differs from traditional geopolitical accounts of security derived from ontologies and anthropologies of political subjectivity, the chapter also clarifies why Foucault concludes that biopolitics simply is a collection of *dispositifs de sécurité*. Strictly speaking, therefore, there is no biopolitics which is not simultaneously also a security apparatus. There is no biopolitics of this, or a biopolitics of that. When one says biopolitics one says security, albeit in a certain way.

Biopolitics arises at the beginning of the modern age, but it does not spring fully formed at its beginning. It would run entirely counter to Foucault's approach to the analytic of power relations to pretend otherwise. While acknowledging a certain kind of precursor in the pastoral power of the Church, with which it appears superficially similar but from which it diverges in its specificities, what Foucault begins to draw out is the logic of formation which takes hold when power takes modern accounts and positivistic enframing of species life as its referent object, and the securing of species life becomes the vocation of a novel and emerging set of discursive formations of power/knowledge. This biopolitical logic of formation also expresses a new and emergent experience of the real. A logic of formation is therefore historical, local and particular. It also installs an ontopolitics as it experiments with novel ensembles or technologies of social practice. However generalised it may become, biopolitics is not itself a universal phenomenon. It is the actualisation, instead, of a specific historical and evolving economy of power relations.

Such ensembles of practices do not actualise themselves in perfect realisation of their logic. First, because their logic is always a contested epistemic object for them. Second, because things always change in unintended ways. Biopolitical security practices do not articulate a design in nature. They are contingent achievements reflecting the partial realisation of designs which seek to enact 'natures'. In the process, there are slippages and breakages, shifts and revisions, for which the original drivers and concerns of biopolitics no longer account. There is nothing unusual in this. It would be unusual if it did not happen. Mutation of the biopolitical order of power relations has continued to follow transformations in the changing order, and ordering, of (living) things. Such mutation has not merely entailed a change at the level of practice. Any change in practice is simultaneously also accompanied by a change in the experience of the real. In general terms the shift in the nature of the real associated with biopolitics, now, is captured by the term 'emergence'.

It is part of our political discourse to talk in terms of a people, a public, a nation or a state as agents having intentions and expressing a view. The life of species being, which Foucault first interrogated in his analytic of biopower, was not this life. Empirically speaking, the life which Foucault first interrogated when inaugurating this analytic of the bio-economy of power relations was that of 'population'. A population is not a subject, a people or a public. A population is a cohort of biological individuals. Specifically, from an insurer's viewpoint, for example, a population is simply a risk pool and risk is not simply an occasion of danger, it is also an opportunity to profit.

Populations are therefore said to display behavioural characteristics and correlations (Hardt 2004: 427). The epistemologies of political subjectivity – especially in relation to traditional security discourses – are preoccupied with establishing secure knowledge about more or less rational choice, interests, intentions and capabilities, and so on. Even when they cannot realise it, which is always, their regulative epistemological ideal is the establishment of causal law. Conversely, the epistemologies associated with the biopoliticised securing of populations are those concerned with surveillance and the accumulation and analysis of big data sets concerning behaviour, the patterns which behaviour displays and the profiling of individuals within the population. Instead of causal law, such power/knowledge is very much more concerned to establish profiles, patterns and probabilities. Here is an illustration.

On the morning of his 65th birthday, the first birthday card which Michael Dillon opened was from his local Sainsbury's store. It congratulated him on reaching his ripe old age and invited him back into the store to collect a free bottle of his favourite Pinot Noir. Anybody who has ever used a store card (in this case a 'Nectar' card) will therefore understand exactly what was going on. Michael Dillon was being biopolitically 'secured' as a 'customer' rather than juridically secured as a subject. He was being governed through his contingent economic freedom to purchase.

Biopolitics is not only a biopolitics of security. The principle of security which operates biopolitically is fundamentally that of economy. Being secured as a customer is one of the very many ways in which biopoliticised security practices now saturate our everyday lives. Most people subject themselves to such everyday security practices most of the time through these and other allied devices; such as, for example, insurance, pensions, investments, securities and risk (Ericson, Doyle, and Barry 2003; O'Malley 2004; Daston 1988; Lobo-Guerrero 2011, 2012). In one sense, therefore, security is a commodity or even a portfolio of commodities. But, more profoundly, security is now the range of technologies and power/knowledge epistemologies which regulate freedom as contingency through the principle of economy. The kinds of means by which Michael Dillon was secured as a Sainsbury's customer are now integral to the biopolitics of security of all modern states, to the operation of the international political economy and to the very 'social inclusion' of individuals in the local economic life of their communities.

This explains the epigraph which heads this chapter. Security and freedom are at odds only for those committed to the idea that civil society and its freedoms precede the historical transformations which created them, as a correlate of the arts of liberal government. For Foucault, civil society was the object domain which the arts of government arising in the eighteenth century posited as the 'natural' limit to the power of the sovereign. In global liberal politics, security and freedom are not rivals. Global liberal politics governs through freedom, fundamentally understood as contingency, and governing through freedom as contingency requires governing through technologies which seek not merely to command but also to profit from contingency. These technologies – as the use of the term 'securities' in global finance for example indicates – are security technologies. What distinguishes them from traditional understanding of security technologies is that traditional security technologies are essentially prophylactic measures, trading in the protection of pre-formed bodies, whereas biopolitical security technologies are regulatory measures, trading in the contingency and futures of bodies continuously in-formation or what nowadays is called 'emergent'. This is not only an extension of the domain of security. It also represents a profound ontopolitical shift in the very character and correlation of governance, freedom and security as such.

Here, life is distributed, weighed and valued, across a shifting terrain of contingent formation rather identity expression or even identity construction. From a simple constructionist perspective, for example, identity has in effect to be written. From a biopolitical perspective, contingent and continuously emergent formations have to be underwritten through a whole variety of calculative practices, not least insurance and those devices which financial markets call securities. Securitising financially has nothing to do, for example, with making an issue a matter of national or international security in a traditional sense. It is not immediately or directly to do with the politics of identity either. Securitising financially is a classic biopolitical strategy which capitalises Life by translating contingency into risk and risk into a tradable asset. Such may become a geopolitical as well as a biopolitical issue, since, of course, it impacts directly on the viability of sovereign states. But its logics of formation lie elsewhere. Biopolitics is therefore not a politics of identity – enacting a self–other dialectic through discursive practices of identity production. It is a complex array of changing mechanisms concerned with regulating the contingent economy of species life. Identity – more precisely subjectification – may follow from this, but identity production is not its original driver.

As we shall see, Foucault identifies 'circulation' as the space of operation of biopolitics. Like all biopolitically driven institutions, therefore, in its preoccupation with the pursuit of economy and profit, Michael Dillon's Sainsbury's store is integrally also concerned with circulation. Circulation is concerned with flows, but flows have to be monitored and regulated. Amongst many other effects, the task of monitoring and regulating flows changes the basic routines and practices of governing institutions including, for example, those of borders. Borders have never simply differentiated inside from outside. (It is common, of course, to observe that borders join as well as separate (Migdal 2004).) In the world of circulation, however, borders are portals. Sainsbury's borders are its check-outs. They are the portals through which access to their goods and services is both granted, regulated and monitored. One of the many purposes of check-outs is to profile and measure the behaviour of that cohort of consuming individuals which Sainsbury calls its customers. Operating in cognate ways, state borders now also deploy the same kinds of surveillance technologies to do similar things. They do not prophylactically secure the state by throwing up a barrier around it. Neither do they constitute an identity for it (Campbell 1998). Their task is to regulate the very re-productive powers of the intercourse transacted by, between and through populations.[2]

Such biopolitical intercourse simultaneously both sustains and undermines itself. Air travel circulates disease as well as tourists, commerce and business (WHO 2003). Similarly, the international financial system may be used to sustain terrorist activities as well as industrial and commercial growth (Biersteker and Eckert 2006). Complex national and international infrastructures vital to sustaining the very ebb and flow of biopolitical intercourse are the same mechanisms through which life also threatens itself. Such dangers may not be posed by a malign agent. It is commonly recognised now that they can be a direct function also of the very complex dynamics of the systems themselves (Perrow 1999). The war on terror has intensified and amplified these characteristic features of the biopolitics of security, but it did not initiate or invent them. It baroquely proliferated them in a spiralling gyre of security and surveillance technologies. The security problematic posed by biopolitics is, therefore, not that of violent death at the hands of other men. It is how to 'make life live'. But what has been happening to life?

This chapter, therefore, asks a second fundamental question: What happens to biopolitics of security when their referent object – life as species existence – undergoes profound transformation and change? I have already begun to signal what asking this second question does. Asking that question takes the analysis of biopolitics of security beyond the terms and circumstances through which Foucault first interrogated it. The question helps revise Foucault's account of the biopolitics of security, and extend its theorisation to the analysis of the political rationalities and technologies of security which are now common among regimes of biopower which increasingly characterise the twenty-first century. At this point, the chapter also identifies three critically important developments; developments which are continuing to have a profound impact on the changing character of Life as the referent object of the biopolitics of security in the twenty-first century. The first is demographic and concerns population. The second is molecular and concerns organic life. The third is digital. It concerns machinic and virtual life. These last two are generically concerned with what might be called the changing vital signs of life and the question of animation – assemblages that display life-like properties. For reasons of space, and because the literature on digitalisation, and cyborgisation, for example, is already very extensive, I confine myself to offering some discussion of the first two only (see Doyle 2003; Thacker 2004; Hayles 1999).

Modern populations, for example, have long displayed what many demographers and health specialists now call 'the compression of morbidity'. What that refers to is the way in which illness and death among Western populations is no longer primarily caused by infectious diseases but is now regarded as a function of the risk factors integral to their very modes of existence. If the so-called 'compression of morbidity' refers to the transformation of populations, the molecularisation of biology has profoundly transformed the very modern understanding of Life itself. The chapter concludes its revision of the analytic of the biopolitics of security by arguing that biopolitics of security in the twenty-first century are best conceived as a recombinant biopolitics of what the molecular life sciences would call 'pluripotent' life.

One final caveat also seems to be required whenever the question of Foucault, and the analytics of power, arises. Foucault never offered, or claimed to offer, a general theory of power. He was quite explicit about this. The assemblages of power which he analysed were those modern assemblages of power which emerged in, and out of, Western Europe from the seventeenth century onwards. These were concerned specifically with the development of modern forms of knowledge and modern forms of freedom. Many other modern ways of exercising power and domination, many other modern ways of being free, were pioneered within and exercised by North Atlantic societies. He never claimed differently. The development of changing formations of power and domination is not confined to the North Atlantic basin either, of course, or to modernity. Foucault never disputed these points – neither does this chapter.

His was instead a novel historical positivity of power relations with a preference for describing mechanisms and interrogating their power effects before generalising principles; although he was not averse to doing this as well, by way of summary account and dramatising operational differences in the many modalities that excite and circulate power. In the process, Foucault constantly emphasised the summary and qualified, rather than definitive, nature of his interrogation of the manifold of power relations which characterise political modernity, and its characteristic modes of governance and rule, including that especially of liberal governmentality (Burchell, Gordon and Miller 1991; Barry, Osborne and Rose 1996; Larner and

Walters 2004; Dillon 2004). One gets the decided impression that a formation of power was for Foucault always in formation, never a final accomplishment. I try to bring that appreciation to my own interrogation of the changing biopolitics of the twenty-first century, and its newly emergent recombinant biopolitical economy of security in particular.

Conceived as evolving through the indefinite finitisations of a force field of time, in which individuation continuously takes shape and transformation occurs, biological entities understood in these terms will not be secured through the practices traditionally favoured by the geopolitical discourses of political subjectivity. They cannot be secured in such ways because, representing a differently understood referent object of both freedom and power, living entities pose a quite different kind of security problematic. For one thing their very presence is not fixed. For another it is their very fructification that is being sought, not their simple protection. In order for living entities to fructify they have to be allowed to do so, or rather, their aptitude for life cultivated through the management of their vital signs of life. Their biological freedom to adapt and change is integral to securing their very existence and the realisation of its potentialities. Securing here is not a condition of possibility of their freedom. It is that regulation of their behaviour which promotes – trades in and trades off – their vital signs of life; specifically these days those concerned with adaptation and transformation. A fundamental paradox thus arises for the biopolitics of security. In order for a living thing understood this way to be secured it has to be allowed – indeed encouraged – to pass out of phase with itself and become something other than what it was, in order to continue to live. That is why biopolitics is not initially a politics of identity or political subjectivity. Identity or political subjectivity is not at first the name of the game, which is not to deny that biopolitical security practices are mined as a rich source of novel subject positions; the additional game of applying the cogito-effect to emergence.

That which endures by moving out of phase with itself, as it transacts complex negotiations with its environment or milieu, is not amenable, however, to being secured by inscription within the boundaries of an identity, a territory or the cogito. Its vital signs of life are precisely those governed by a different operational dynamic of the contingent. It is this which typically invokes biopolitical security practices. The name of the game here is 'capacities for adaptation and change', sometimes known as 'fitness'. It is that which ultimately promotes the very understanding and command of the dynamics of hetero-genesis into the empirical referent of the biopolitics of security of the twenty-first century.

The biopolitics of population: From '*Le Genre Humain*' to '*Être Biologique*'

When, according to Foucault, modern political power first assigned itself the task of 'administering life' its governing technologies revolved around two poles of life. The one centred on 'the body as a machine: its disciplining, the optimisation of its capabilities, and the extortion of its forces' (Foucault 1998: 139). Foucault called this the 'anatamo-politics of the human body'. The second formed somewhat later. It focused on 'the species body, the body imbued with the mechanics of life and serving as the basis of the biological processes: propagation, births and mortality, the level of health, life expectancy and longevity, with all the conditions that can cause these to vary'. These regulatory mechanisms constituted 'a bio-politics of the population' (Foucault 1998: 139). These two new modes of governmentality were not antithetical: 'they constituted rather two modes of development linked together by a whole intermediary cluster of relations'. Here, then, as modern power learnt, in its scientific ways, what it meant to be a living species in a living world – 'to have a body, conditions of existence, probabilities of life, an individual and collective welfare, forces that could be modified and a space in which they could be distributed in an optimal manner' – biological existence was reflected in political existence and power, not simply dealing in legal subjects but with living beings, newly understood in these increasingly finitudinal ways, took 'life and its mechanisms into the realm of explicit calculations and made knowledge/power an agent of the transformation of human life' (Foucault 1998: 143). Biopower in particular, Foucault was to elaborate in his lectures, 'deals with the population, with the population as a political problem, as a problem which is at once scientific and political, as a biological problem and as a power's problem' (Foucault 2003: 245; see also Elden 2006).

What Foucault initiated in his analytic of biopower, and what follows here, is not an espousal, however, of the biopolitics of security.[3] Neither is it an argument to the effect that all other politics of security and freedom have been displaced by the biopolitics of security. What Foucault originated, and what this chapter seeks to theorise in a Foucauldean vein beyond Foucault, is an 'analytic' of the biopolitics of security. It is perhaps helpful, then, to state briefly what is meant by an analytic

An analytic of power is an exercise of thought concerned with detailing the operational logics, forces and dynamics at play in a specific configuration of power relations. One might say, it is as concerned with design principles as much as outcomes and effects. Since power for Foucault is always a matter of the relations established by a set of practices, specifically those concerned with changing mechanisms of power/knowledge, a Foucauldean analytic of power is necessarily also an exploration of the forces, logics and dynamics at play in assemblages of such practices. It is important also to explain that for Foucault, as much as for this book, every assemblage of such practices is an assemblage of 'truth-telling' which not only seeks to regulate life but also reflects and enacts a certain understanding of 'the real' as it does so. In the process what Foucault calls a grid of intelligibility is established, which grids of intelligibility simultaneously also become the means by which life is weighed, distributed, valued and found wanting, as well as good. A grid of intelligibility is in short an accounting and a valuing machine (Foucault 1989: 422; see also Elden 2006, 2007).

Such modern assemblages of power/knowledge will, according to Foucault, also critically interrogate the very ontology to which they give expression. Intelligently remedial in relation to the governmental technologies which they employ, biopolitical security practices simultaneously also explore how 'the real' they presuppose might be better understood and acknowledged. The biopolitical understanding of the real – its experience of the 'real' of living things – is baroquely played back into refining the technologies designed to regulate and capitalise upon it. It is for these, among other reasons, that Foucault referred to the regional ontologies employed by such truth-telling practices, for example those of 'Life', 'Labour' and 'Language' in *The Order of Things*, as 'quasi-transcendentals' (Foucault 1989: ch. 8). Although Foucault does not directly address them in this way, 'the event', 'the contingent', 'population' and, as we will see, 'circulation', do all figure more or less prominently as such in his account of biopolitics. Thus: 'Despite its apparent and violent naturalism, what the baroque offers us never remains a pure and simple realism' (Maravall 1986: 257).

Foucault first taught that the biopolitical promotion of human being as species existence originally developed around the interrogation, classification and statistical analysis of population in the burgeoning towns and cities of France in early modern Europe. Here, he says, a new political object of governance, a new reality, emerged; that of the population rather than the people.

The life of species being is, however, neither single nor fixed. On the one hand, biopolitics is critically dependent, epistemically and ontologically, on what the sciences of life say that species life is, what it consists in, what its thresholds are, and on the mechanisms that these make available for intervening into living processes in order to further what Foucault taught was the vocation of biopolitics; the promotion of the productivity and potentiality of species life, abstracting its very vital signs of life the better to utilise them in the further promotion and exploitation of life's potential. Biopolitics is similarly also susceptible to historical transformations in the very character of the populations which it seeks to regulate as well as in the scientific understanding of the very nature of living material as such. But what has to be understood, first, is a base distinction drawn within the category Life itself.

Intrigued by the mechanisms through which the basic biological features of the human species became the object of political strategy, when Foucault first came to interrogate biopolitics he noted how it differentiated life as species existence from life understood in other ways. Life understood as species existence – *espèce humaine* – differs fundamentally, for example, from life understood as '*le genre humain*' (Foucault 2004b). The root of *le genre humain* – *gens* – refers to the *jus gentium* of Roman and medieval law. Usually translated as 'the law of nations', it was extensively treated in the work of two early modern international jurists, Hugo Grotius and Emmerich de Vattel (Grotius 1925; Vattel 1883). Important to the development of early modern national and international understandings of rule and power, the *gentium* of *jus gentium* invokes the juridico-political and cultural notion of a 'people' or 'peoples' belonging together in respect of law and custom, not the biological notion of 'species' (the root of *espèce*, or *être biologique*), in which the principle of belonging together is furnished by shared biological properties. The move from *gentium* to *espèce* thus effects a transformation in the very understanding of what it is to be a living being, and, correspondingly, of the governmental regulation of such a living thing. What followed in that shift from *gentium* to *espèce* was a transformation not only in the referent object of power relations but also in the very mechanisms by means of which power operates and circulates. Specifically, in relation to *espèce humaine*, power comes to be exercised on, in and through the biological mass which constitutes the 'species', rather than the juridico-political and cultural processes of belonging and rule said to constitute the *gens* of *gentium*, or of '*le genre humain*' (see Bartelson 1995).

Dispositifs de Sécurité

Foucault began a lecture series entitled *Security, Territory, Population*, on 11 January 1978. Announcing the title of his lectures, and cautioning that while his enquiry would focus on the set of mechanisms through which the basic biological features of the human species became the object of political strategisation he was not concerned with a general theory of power, Foucault first posed the obvious question: What are we to understand by 'security'? His answer was an unusual one. He offered no general theory of the universal value of security. Nor did he proclaim it as a general necessity for life. Typically, he began instead to talk about certain specific mechanisms which emerged at the start of the eighteenth century. We have to reconstruct Foucault's understanding of the biopolitics of security both from his account of these technologies, and from his more general understanding of the process of problematisation (Foucault 1988, 1994, 1996).

From a Foucauldean perspective, before security can become a concept, a value or a valuing process it must first be problematised. Consequently, different problematisations of security will be comprised of different discourses of danger. Different discourses of danger will revolve around different referent objects of security, such that different referent objects of security will give rise to different kinds of governmental technologies and political rationalities. Such problematisations will be derived from the operation of specific complexes of power/knowledge. Different assemblages of power/knowledge will themselves also be comprised of different mechanisms, techniques, instrumentalities, rationalities and discursive formations. Hence the overall baroque character of our politics of security. The problematic of security posed by Life, for example, will not be the same as that posed by the 'human' (*anthropos*), by political territoriality (*la patrie*, motherland, fatherland), or by sovereignty (*Volk*, *Reich*, *Führer*, people, or *demos*). Neither will the security apparatuses that develop around these different referent objects.

Modern times have been distinguished by at least two great problematisations of security. Foucault retells the story of the one (geopolitics) as he introduces his novel account of the other (biopolitics).[4] The first, revolving around the referent object of sovereign territoriality has been that traditional geopolitical narrative of security which emerged with the establishment of the European state system following the Peace of Augsburg and the Treaty of Westphalia. This military strategic international relations discourse came to dominate our modern discourses of security. It articulates what amounts to a needs or value hierarchy in which security is posited as a universal value that is foundational to the very possibility of individual and collective life. Politics emerges from a social contract, founded in this needs hierarchy. It takes the form of a protection racket in which loyalty to state authority is derived from the threat of insecurity said to follow from not having the protection of the state; which apparatuses, in practice, do most of the threatening. Consider the way in which the US/UK invasion of Iraq is now said to have stimulated the security threat posed by radicalised Muslim youth to the UK in particular. The state must therefore always secure its monopoly of the legitimate use of force when posing as the primary security provider. In order to do so it must similarly also monopolise the legitimate definition of threat on the basis of which its claim to monopolise the legitimate use of force is ultimately grounded. Such an account of security has ramified into a whole variety of subsidiary concerns including, for example, civil/military relations; the relative mix of military and other forces which state power (hard and soft) must deploy; the impact of the emergence of international and non-state actors on the state security system; and an irresolvable 'level of analysis' debate concerning the radically undecidable relation between individual and collective security.

Every traditional geopolitical discourse of security invokes security as the limit condition. War in particular operates as the privileged locus of the real for traditional security analysis; the reality said to trump all other realities in the hierarchy of values and needs that underpin its traditional geo-strategic discourses. War here is, of course, construed as one of the instrumental means available to the state as actor. In an earlier lecture series pursuing these themes, however, Foucault controversially reverses Clausewitz's dictum that war is the extension of politics by other means (Foucault 2003). He argues instead that war is not an instrument available to a political subject, but a grid of intelligibility from which modern accounts of liberal political subjectivity, in particular, arise. One might therefore gloss Foucault, here, and say that security discourse is the *logos* of war expressed as a *logos* of peace.

Thus, while Foucault's analytic of the biopolitics of security typically focuses on micro practices it is important to underline that this analytic articulates a qualitatively different problematisation of security than that which characterises geopolitics. The distinction is not, in short, founded on the basis of a naive level of analysis distinction between macro and micro practices. These two security analytics do not inhabit the same plane of analysis, albeit they coexist and continue to correlate throughout the history of political modernity.

It is nonetheless difficult, however, to sustain the privileging of the micro over the macro which Foucault ordinarily champions in his various analytics of power, because the problematisation of security provides such a powerful point of intersection between geopolitical and biopolitical security analysis. With discourses of security, the micro-political regularly becomes the macro-political; when minor infringements of codes of normalisation, to do especially these days with religious dress codes for example, become incitements to political and nationalistic fervour, and when macro-political calls to nation cash-out into the detailed inscription of everyday conduct. This point was noted by Colin Gordon, for example when he referred to security as a 'specific principle of political method and practice' which joins the governing of the social body to 'proper conduct' of the individual, to the governing of oneself. Foucault also referred to it as the process of internal colonisation (Gordon 1991: 20). There is thus no geopolitics that does not imply a correlate biopolitics, and no biopolitics without its corresponding geopolitics. Giorgio Agamben states this point succinctly, in his rival account of biopolitics: 'It can even be said that the production of a biopolitical body is the original activity of sovereign power' (Agamben 1998: 6).

Take, for example also, the ways in which Malthus' population science was recruited into racial geopolitical projects in the nineteenth century. Consider how racialist doctrines informed European imperialisms (liberal as well as autocratic). Note, thirdly, how racial 'science' informed German geopolitics throughout the first half of the twentieth century. Spatiality was directly linked there to race through '*Raum*' and '*Lebensraum*'; where '*Raum*' constituted the very geopolitical project required for species existence (Ó Tuathail 1996). Within liberal regimes of power today, molecular biopolitics is now also being invoked to encourage the return of eugenics (Agar 2004; Mills 2007).

What is common throughout traditional geopolitical security discourse is, however, what we might best call a prophylactic interpretation of the security problematic. Security is essentially posed as a protective or preservative measure thrown around a valued subject or object, which subject or object is presumed to exist prior to, and independently of, the security practices which claim to act in their welfare. The entire edifice of modern war, diplomacy, *Macht*, *Realpolitik* and *raison d'état* was raised and continues to function on this basis, with significant revisions and additions following the rise of liberal internationalism in the twentieth century. Students of international relations have long noted, however, that security practices themselves are instrumental in constituting the very subjects and objects which they claim to secure (classically, Campbell 1998). It is evident also, of course, that security cannot simply be an assemblage of protective devices since you cannot secure anything without first knowing what it is that you are securing. Even traditional geopolitical security mechanisms must therefore also be comprised of complex epistemic schemas as well. Thus geopolitical power/knowledge is not only concerned with the instruments of statecraft – *arcana imperii* – it has also to concern itself with making the very subject or object to be secured transparent to knowledge as well: hence, its allied and expanding concern with culture and identity, as well as science and technology, for example, in addition to the early biopolitical technologies of statistics and logistics (Petty 1690).

The aleatory biopolitical economy of population

Biopolitics of security are, thus, different from rationally conceived geopolitical security practices. Whereas national and international law, for example, proscribes certain behaviour, and carceral institutions install panoptical systems of disciplinary surveillance in the execution of the correction prescribed by law, Foucault draws attention to other mechanisms that were being extensively developed, especially from the beginning of the eighteenth century onwards. These were concerned neither with the practices of the law nor the disciplining of bodies. They were concerned instead with the distribution and statistical patterning of behaviour. Typically they began by asking, for example: What is the average rate of criminality for a certain type of human being – male/female, young/old – in a given town or region, in respect of certain crimes and so on? What other factors might be correlated with the incidence of such behaviour including, for example, social crises, famines or wars? What factors correlate in particular with the rise and fall of the incidence of such conduct (Foucault 2004b: 1–4)?

We might say that such mechanisms were generically concerned with mapping the incidence of contingent behaviour and correlating it with a whole variety of environmental and biological factors. These practices, Foucault says, were not characteristic either of the legal code or of disciplinary power mechanisms (Foucault 2004b: 1–2). Instead they constituted what he called apparatuses or *dispositifs* of security. Foucault did not claim that these devices were entirely novel. They were not. Neither did they cancel out geopolitical sovereign juridical and disciplinary forms of power. He observes instead that while the progressive expansion of legislative measures, decrees, regulations and circulars fostered the deployment of these mechanisms of security – since their introduction introduced more 'criminality' and therefore offered an incitement to its biopolitical analysis – biopolitical security mechanisms increasingly also relied upon data derived from the intensification and extension of increasingly novel forms of counting, accounting and surveillance. These were stimulated as much by civil and commercial as they were by juridical or military-logistical concerns.

These biopolitical security mechanisms were, therefore, not confined to documenting the contingent correlations of criminality and its repression. Since they revolved around the advent of power relations which took the life of species being as their referent object, they were generically concerned with species being and, therefore, also with economy, health and medicine as well. Indeed, the biopolitical problematisation of security revolved around economy as it also introduced a general biologisation of security discourses; including health and medicine, especially, within the ambit of security thinking. This development was to become widespread by the beginning of the twenty-first century. It is not confined to the so-called civil sectors of society either. It is now also common throughout much contemporary military strategic discourse. The current 'Revolution in Military Affairs' (RMA) is, for example, not simply a Military Technical Revolution characterised by the informationalisation of weapon systems and military force structures, combined with the correlate weaponisation also of information (Dillon and Reid 2009). A significant transformation in the 'equipmentality' of security and war, to use a Heideggerian expression, it is very much also a transformation in military strategic cognition throughout which biologised thinking, together with health and medical analogies, abound (Heidegger 1962: 559).[5]

Thus, Foucault explains how, whereas disciplinary power/knowledge enframed the danger posed by plague in terms of the practices of confinement, at the beginning of the eighteenth century security practices began to address smallpox and the threats posed by other infectious diseases in a quite different, biopolitical, way. The security problematic which these posed was addressed less in terms of exclusions than in terms of how to fashion modulated and targeted strategies for combating endemic diseases; which strategies derived from the analysis of the contingent distribution patterns which epidemics were found to display as they coursed their ways through populations (Foucault 2004b: 1–2).

Such biopolitical mechanisms or security technologies were characterised by a number of distinguishing features. First and foremost they were preoccupied with the aleatory. They were in other words concerned with the fundamentally contingent character of behaviour and the contingent correlation of behaviour with an (in principle) infinite variety of correlate factors (Foucault 2004b: 1–2). The second was a preoccupation with economy (Foucault 2004a, 2004b). Seeking to map the correlations of contingent behaviour, these mechanisms were concerned with the cost of behaviour; in particular asking, for example, if the cost of correction exceeded the cost of the criminality. From that followed an additional concern with applying cost–benefit calculus to different polices and with the modulation of such policies looking to establish a band-width of the acceptable. Basically, Foucault says, the fundamental question was economic; the economic relation, for example, between the cost of repression and the cost of delinquency.

He then went on to itemise a variety of other features. In addition to their preoccupation with the aleatory and the economic, biopolitical security technologies were also distinguished by what Foucault called their space of operation (see Casey 1997; Elden 2001). If economy was one of the emergent quasi-transcendentals of biopolitics, its space of operation was not defined by the usual geo-economic preoccupations with global regions, national territories, strategic resources and strategic cities or other sites of military strategic and geo-economic significance. At first, he says, the space of operations characteristic of biopolitical security technologies was that provided by 'population'; where population served in effect as both datum and source of additional data. 'Population' thus came to offer a new reality to power/knowledge. Security mechanisms, as Foucault described them here, statistically mapped the contingent behavioural characteristics of populations. They did so guided by some basic questions of economy: distribution and cost–benefit. They also acknowledged a certain material independence of this space of operation from the techniques which began to interrogate it. Population and its sub-fields were not only accorded an objective status by the very 'sciences' which gave them specificity, their associated biopolitical security mechanisms encouraged regulatory policies informed by what was made known through the scientific interrogation of the independent dynamics, laws and regularities which populations displayed; and which population science sought to make more widely available.

As the *Security, Territory, Population* lectures in particular advanced, however, it nonetheless also became clear that population was for Foucault embedded in a much wider and more generic space of operation. Foucault simply called this space of operation, 'circulation' – circulation in the widest and most generic sense of the term. Within that generic concern with circulation, biopolitical security mechanisms also concerned themselves with the incidence of possible 'events'. In that they nonetheless also began to tell a tale about the real: that the real made available through the optic of living things existed in a different temporality even from that of an open-ended history. The reality of time, the time of the real was [is] 'evental'. That is why events come to be understood biopolitically not only as the temporal uncertainties which occur in and through the very forces and conduits which comprise the circulation of species life; the events which insurers, for example, underwrite or the events with which anti-terror, homeland security and national resilience experts seek to deal. These may be things that one will want to prevent before they happen. The issue to which Foucault gave most historical attention was the one which preoccupied early modern technologies of governance through *polizeiwissenschaft*. This was famine or 'dearth' (*la disette*). Such events may also be things one might want to engineer, such as prosperity. Events, however, reflect the very 'eventalness', contingent historical 'thrownness', of species existence as such. Hence the Being of species being is complexly bound up with varying accounts of the philosophy of the event.[6]

The general problematic with which these biopolitical security techniques of population became preoccupied was thus the regulation of circulation, including not simply, for example, the freeing of circulation and the unblocking of log-jams, but more generically also with security the problem of differentiating good circulation from bad circulation. Free circulation came to be generally favoured as laissez-faire and the ideas of the economists and physiocrats took over from those of 'police' and the mercantilists – specifically in regard, for example, to the free circulation of grain through market mechanisms in order to avert famine. But the security problematic posed by the regulation of species existence in these specific ways was not simply that of maximising the liquidity, or regulating the viscosity, of circulation. Here the ambiguous, or dual, nature of circulation as the space of operation for biopoliticised security practices emerges most clearly.

While in their emergent preoccupation with the economy of contingent distributions, biopolitical security techniques were generally well-disposed towards laissez-faire, effective regulation of laissez-faire nonetheless also required an increasingly sophisticated knowledge of circulation, if subtly crafted interventions were to profit from the force of the independent dynamics of circulation, contingency and economy – in the form, for example, of markets. What was required were the means by which circulation, which was desired, could be distinguished from circulation which was not. In effect, the security problematic of circulation as the generic space of operations for these new biopolitical security mechanisms posed itself in terms of differentiating good circulation from bad circulation. It also posed itself in terms of the balance to be struck between too little and too much regulation of the manifold circuits of interchange that characterise the life of species existence. These dilemmas were to become the classic dilemmas of an expanding system of biopower relations distinguished by the manifold ways in which it encouraged biopolitical self-governance, via contingency management, as the principal means both of securing the welfare of populations and their everyday self-rule. (It will not escape the attention of those documenting the war on terror and homeland security and national resilience strategies that these all closely revolve around such a way of problematising security (see Dillon 2005).)

It is also important to underline what distinguishes 'economy' here in Foucault's account – something which preoccupies *The Birth of Biopolitics* lectures in particular (2004a). Foucault does not approach the problematic of economy by positing the existence of an economic subject, distinguished by rationally calculated utility. Once more he attends first to specific practices, such as those which emerged to deal, for example, with the problem of famine. Scarcity – in effect a quasi transcendental for capitalist economics – emerges for Foucault against a more ancient background – that of 'fortune' – in which man's evil also figured. As Foucault explained, fortune – good or bad – was never simply an acknowledgement of impotence. Every rendition of uncertainty, or accident, takes place against some account of the true and the real. (Risk discourse is a veridical apparatus.) Within each account, provision is also always made for some means of addressing chance, including that of seeking some measure of redress for its effects (classically, see Machiavelli 2005). Chance is therefore always a political and moral concept which arises in the context of changing cosmologies of being. However chance is figured, it is always figured as an integral part of some political and moral economy. As the modern developed its increasingly calculative ontopolitics of existence, its account of fortune progressively became a function instead of the kind of statistical security techniques which Foucault discusses in these lectures – namely those concerned with the regulation of behaviour through the analytic of its contingent distribution. Fortuna, chance and destiny became contingency. With the advent of more precise mathematical techniques, specifically probability, contingency was increasingly commodified. Contingency commodified became 'risk'. Economy seems then to derive for Foucault from an originary preoccupation with the specificities of circulation, distribution and contingency management (Foucault 2004a, 2004b). A second feature which distinguishes the way economy plays out in Foucault's analytic of biopolitics is the emphasis on fungibility – the infinite substitutability of one thing for another in the circulation or intercourse of living things made available by the advance of the calculability of circulation as such.

In ways which anticipate later modern accounts of the function of global cities, the early modern town, for example, was conceived by biopolitical security technologies as a node in a network of many circulatory connections and practices subject to a plurality of possible correlations and events (Sassen 1991). Species life, in general, thus emerges as a manifold in Foucault's account, and its biopolitics of security deal with the contingent multiplicities engendered not only by the way that species life circulates, but also by the manifold of things, such as disease for example, which, in addition, circulate through life. Every biopolitics of security, we might therefore now add, has an allied microbiological politics of all of the things circulated in its turn by life.[7] Security, Foucault therefore says, 'involves organizing, or anyway allowing the development of ever wide circuits' (Foucault 2004b: 70): patterning and probabilistic analysis of the incidence of contingent events, and the aleatory distribution of contingent behaviour among populations gathered pace.

As this story proceeds, in particular through the first half of the *Security, Territory, Population* lectures, Foucault tracks its emergence also through the economic debates which occurred between the mercantilists and the physiocrats which preceded the emergence of liberal economic thought, and the advent, in addition to 'population', of another new social reality: that of the notion of 'the economy' itself, and of its market mechanisms. Key to these political economy debates was of course the question of freedom (Foucault 2004a).

Whereas the basic function of discipline was to prevent everything 'even and above all the detail', the function of security was to rely on details that were not intrinsically good or bad but necessary; inevitable, as with so-called natural processes; crucially those of contingently correlated circuits of production and reproduction, economic and sexual, micro as well as macro-biological, material as immaterial. And while such details may not be pertinent in themselves, they may be combined in ways that deliver up pertinent information at the general level of aggregation associated, for example, with population. Biopolitical security apparatuses did not, therefore, side with what was to be prevented or what was to be made obligatory by law. Their referent object was instead a new 'reality'; one increasingly understood to be the very contingent taking place of factically finite things. The socio-technical mechanisms which began taking this reality as their referent epistemic object could nonetheless also be put at the disposal of sovereigns; who, of course, frequently sought to finance them. Evaluating also whether or not such eventalness was desirable, biopolitical security mechanisms were designed to make things operate by regulating how their aleatory features contingently correlated. One might now say that biopolitics of security emerged as an early form of so-called 'effects' and 'evidence-based' operation.

This very tight correlation of biopolitical security apparatuses with contingency, eventalness, economy and circulation is, Foucault also says, profoundly linked to the rise of liberalism as a regime of power relations (the general thematic of *The Birth of Biopolitics*); acting so that reality develops, is free to go its own way, follow its own course according to laws and dynamics amenable to epistemic interrogation and thus ultimately also to modulated policy intervention. In short, security was fundamentally correlated with a certain understanding of freedom as well. This freedom, he was careful to gloss, is not concerned with the exemptions and privileges that attach to a person. It concerned instead the very possibility of movement, change of place, and so on. Above all, and in summary form, it was the freedom of circulation, and the freedom to circulate, in the very broadest sense of the term. Abstracting further from Foucault's account, one might therefore say that the problematic of biopolitical security apparatuses is fundamentally that of securing the contingent freedom of circulation. That is how and why we are now able to understand the epigraph which heads this chapter: biopolitically, 'freedom is nothing else but the correlative of the deployment of apparatuses of security'.[8]

Peculiarly characteristic, in particular, therefore, of liberal governmental power, power operated here in ways that mimicked currency rather than mechanics (Foucault 2004a). Its logic was similarly more strategic than dialectical. Dialectical logic presupposes contradictory terms that are nonetheless situated within a realm which is ultimately homogeneous. For two propositions to be contradictory or not, it is necessary as, Leibniz says, that they have something in common on the basis of which they can contradict one another. Strategic logic presupposes, instead, that it is possible to connect different terms productively, which terms nonetheless remain disparate (Foucault 2004b). Think, for example, of the disparate terms which risk analysis combines into the mathematical commodification of contingency. Between the disparate, contradiction does not even remain possible; but productive correlation does. The productivity of strategic logic derives from connecting the heterogeneous into different cartographies of the contingent, rather than resolving the contradictory into higher forms of the same. Strategy is thus a typically baroque *ars combinatoria*, presupposing an irredeemably heterogeneous universe, while dialectic is an *ars differentia* presupposing a homogeneous one. Here, eventually, in the contemporary biopolitics of security in particular, the distinction between strategy and tactics becomes otiose, since according to contemporary understanding of what it is to be a living thing, and the art of living it, strategic logic is no longer telic but a re-combinatorial art of baroque design.

Foucault was therefore not only observing a change in the exercise of power and the conduct of conduct, when noting the emergence of the biopoliticality of liberal governmentality. He was also observing how such changes were taking place in correlation with complex changes in truth-telling practices concerning not only freedom but also uncertainty, the nature of the contingent as well, prospectively, as the very understanding of 'species' itself; including species as money, species as generic scientific taxonomy and species as biology (Dillon 2007a, 2007b; Poitras 2000). The space of problematisation of government was, therefore, also related, amongst other factors, to that of the changing problematisation of chance. These developments were in turn related to a changing biopolitical problematisation of both freedom and security in which while the referent empirical object of security was initially population rather than territory, it progressively became that of circulation *tout court*.

The question of contingency, or 'the aleatory', thus arises for Foucault as one of those factical elements, or 'natural' processes, to which liberal governmentality must attend, with which it must deal and in relation to which it has to regulate and evaluate its own performance and effectiveness in its ambition to exercise power over life. In governmental terms, the contingent features which life and populations display are not an ideological disguise for the operation of some hidden interests. Neither are they part of a dialectical historical process. They are a function of truth-telling practices, specifically of the life sciences and of the burgeoning sciences of uncertainty and risk. These perform a whole variety of governmental as well as scientific functions, not least in telling different stories about different categories of living things and their governability, as well as what falls into the category of living thing as such (Dillon 1996). It follows that the changing biopolitical security mechanisms of the biopolitical economy of security are characteristically concerned with the random elements displayed by life in general, and populations in particular.

Foucault concludes by stating that biopolitics *tout court* simply are *dispositifs de sécurité* (Foucault 2004b: 2). Biopolitics of security do not preserve and protect in the classical sense of security. They promote and regulate. The reason is simple. Classical security discourses presume – by one means or another – that they are dealing with a fixed object. Biopolitical security discourses and techniques deal with an object that is continuously undergoing transformation and change through the manifold circuits of production and re-production which comprise the very eventalness of its biological existence. *Biopolitics of security thus secure by instantiating a general economy of the contingent throughout all the processes of circulation which impinge upon the promotion of the re-productive powers and potentials of Life as species being.* That is why we can now say that in an age gone global the biopolitical problematisation of security has become the imperial struggle not simply to seize territory, control resources or even reconfigure state apparatuses – although these remain traditional security concerns – but to secure the changing and manifold processes of global circulation as such.

No politics of life could remain unaffected by the impact which the changing demography of population and the molecularisation of biology have had on life as species being during the course of the twentieth century. Biopolitics has not only, therefore, been transformed by what the life sciences understand a living thing to be, and how life processes can be manipulated by them in order to extend and promote the fructification and vitality of living material, it has similarly also been transformed by what has been happening to the populations of developed societies. Population dynamics and the life sciences are, moreover, complexly related. Across Atlantic societies, these developments are currently revolutionising the biopolitical security technologies of health management as much as they are those of military strategic discourse, the war on terror, homeland security, national resilience strategies and allied corporate security practices (Arquilla and Ronfeldt 1997, 2000; Biersteker and Eckert 2006; Cooper 2006a; Greenberg, Weschler and Wolonsky 2003).

Specifically, the promise of molecular medicine allied to what health professionals call 'the compression of morbidity' among developed populations has already begun to transform the biopolitics of life itself. I deal briefly with the compression of morbidity in the next section and with the molecularisation of biology at greater length in the section after that. Space does not allow a fuller account of either development so our primary concern is to move towards outlining how these developments began to inaugurate a recombinant biopolitics of security, whose generative principle of formation is not simply species being but species being as what molecular biologists would call 'pluripotent' life. Such population dynamics and molecular science are also in the process of being complexly related through the pursuit of biomedicine. Whereas the genetic revolution has received the most publicity, the compression of morbidity has been no less influential at micro and macro levels in transforming the biopolitical rationalities and security technologies of the last century. If there has, then, been a compression of morbidity for populations, medically, there has assuredly also been a compression of life, politically (a conflation of its manifold of existence), and via these and many other mechanism, into our increasingly baroque politics of security as well.

Population: the compression of morbidity

In a now widely celebrated essay first issued as a keynote address to the US Institute of Medicine at the National Academy of Sciences in Washington in 1982, the celebrated American Physician James Fries presented a paper concerning 'the compression of morbidity' which he claimed had transformed morbidity throughout the populations of the United States and Western Europe during the course of the twentieth century. The changes to which he referred, together with the simplicity and eloquence of the thesis that he presented, continue to structure and inform biopolitically focused health policies today. Fries' thesis was as follows.

Western societies entered the twentieth century in an era of infectious disease. In the United States, for example, tuberculosis was the number one killer while smallpox, diphtheria, tetanus and other infectious diseases were prevalent. Curative medicine, allied to other biopolitical 'social' security strategies, transformed the morbidity of Western populations, however, by virtually eliminating such diseases. Consequently, an entirely different demographic of morbidity has emerged. This new landscape is characterised by chronic rather than infectious diseases. Chronic diseases are in turn distinguished by a variety of significantly different features from those which characterised infectious disease. These features include the following:

1 Susceptibility to chronic illness is a universal among Western populations.
2 Chronic illnesses are characterised by early onset.
3 Chronic illnesses are progressive.
4 Such illnesses become clinically explicit only at certain symptom thresholds; angina or heart attack signal the presence, for example, of heart disease that has been developing for many years.
5 Chronic illnesses are multifactorial.
6 Differences between individuals are manifested by the rate at which such diseases progress rather than the bald presence or absence of the disease.
7 The onset and rate of progress of chronic illness are a direct function of exposure to risk factors. These in turn are very much a function of how lives are lived: 'lifestyles'.

Whereas infectious disease invited drug cures, chronic illnesses invite pre-emption and prevention together with medical campaigns aimed at changing lifestyles, backed by widespread use of surveillance (or 'screening' in medical terms) and the statistical documentation of the incidence of disease distribution spatially as well as temporally. All these developments have in turn begun to be combined with the pervasive regulation and management of risk, and the astonishingly greater resolution given to the analytic of risk which resonates throughout all aspects of Western society. Closely allied with the concern for risk factors arising from lifestyles, is an intense individualisation and personalisation of medicine as well. Whereas Foucault spoke of the ways in which liberal biopolitical governmentality encouraged the development of the entrepreneurial self, the compression of morbidity witnesses the emergence of a self entrepreneurially responsibilised to secure its own healthcare (Foucault 2004b).

Such developments similarly reflect the intense preoccupation in military corporate and homeland security domains with the same kinds of preoccupations and analytical devices allying screening and surveillance with distribution-patterning and risk analysis across mass populations locally as well as globally. The biopoliticisation of security discourse in these more traditional areas of security analysis also reflects a massive shift to pre-emptive and preventative strategies. While I do not have the space to trace the connections, the correlation is no mere accident. Granted that both military and corporate strategies are driven by their own local and global dynamics, their informationalisation and biologisation nonetheless also directly reflect these developments as well (Cooper 2006a; Dillon and Reid 2001). The reason for this, I maintain, is that they not only borrow from, and leach into, one another. They share a similar if diffuse account of the 'real'.

Fries also argued, however, that Western societies are entering a new health phase in which the significance of chronic diseases will also decline as these in turn yield to well-targeted health policy interventions. Once such biopolitical strategies take effect, he claims, the onset of chronic illness will be delayed and its progress controlled by advances in molecular medicine and new genetically informed pharmacological regimes. Life expectancy is then expected to rise towards the figure of absolute life span. The duration of chronic illnesses would thus be shortened and death would result from the exhaustion of the reserve capacity of the vital organs rather than the lethality of chronic illness. Ideally, for Fries, people could therefore expect to live – actively and well – until the onset of terminal decline, which would be short and, if not sweet, at least less distressing than years of incapacity. By these means life potential and not just life expectancy would be maximised. That, precisely, is what he meant by 'the compression' of morbidity. From a biopolitical perspective, 'senescence' – the ageing process itself – is already becoming an increasingly important security problematic for biopolitical regulation by actuaries and underwriters, for example, in the classic biopolitical security technology of insurance; as well as for health scientists, health professionals and health managers.

Recombinant biopolitics: securing 'pluripotent life'

As life, and so-called 'lifestyles', change, then, so also will the regime of biopolitics to which life itself is subject. Just as population has changed, so also has life.

In the molecular age, life is no longer simply the life of populations as Foucault documented their emergent biopolitical regulation at the beginning of the eighteenth century. Life in the molecular age already poses a new reality to and through biopolitics, because molecular science has transformed what we understand a living thing to be. It has also revolutionised the very intersections of life and death, including lives on the margin. Lives on the margin in fact become a prevailing concern locally as well as globally, politically as well as medically, socially as well as economically. Marginal life seems no longer 'marginal' but, as with failed states, rogue states, terrorising dissidents coursing through the capillary infrastructures of global society, embryonic life or the material and ethical dilemmas posed by comatosed patients, marginal life emerges as a central organising category in the biopoliticised power relations and continuous policy preoccupation of biopolitical governmentality. Increasingly characterised by a preoccupation with hetero-genesis, recombinant biopolitics is as concerned with pre-life as it is with post-life as it pursues its vocation of making life live.

The molecularisation of life has equally transformed what we are capable of doing to and with living material; the term pluripotent, for example, which we apply to the concept of life so characteristic of the molecular age, is derived from its use by molecular biologists to describe the potentiality of stem cells. Beyond the molecular and into the nano age technoscience is pursuing the possibility of recombining all materiality – organic and inorganic – thorough manipulating it at the level of its common atomic structure. In these and other ways, life itself has come to be profoundly problematised by the arts less of Life than of 'animation'. Recombinant biopolitics is the biopolitics of the molecular age because what matters here is not simply life's molecular structure but the massive advance which has also been made in exposing the mechanisms of hetero-genesis as such, that which in 'truth' makes life live for the molecular age.

Particularly in respect of bioscience and biomedicine, currently now exemplified by the field of regenerative medicine in general and that of tissue engineering and stem cell research in particular, we are entering an age in which scientific knowledge does not merely seek to represent the truth of life (Cooper 2006b; Waldby 2006). Nor is its ambition simply to manipulate life forms. The molecularisation of biology has radically extended our ambitions in respect of species existence: to control and command the morphogenetic process itself. Its ambition is not simply to manipulate specific actualisations of life, but to control the very production of living material irrespective of the specific forms in which it comes. In the process, its rendition of life has also changed. Life becomes equated with its pluripotency, and the object of biopower becomes preoccupied not only with strategies of resilience, self-repair and regeneration but also with instigating new life forms.

Following biological research into stem cells and tissue engineering, for example, life is more characteristically to be understood now as a pluripotent phenomenon. Pluripotent, here, means containing the capacity for regeneration and renewal through processes of self-differentiation via recombination. As Melinda Cooper has observed, operating at the forefront of biomedicine, tissue engineering does not simply concern itself with the transplantation of frozen forms and rigid bodies (Cooper 2008). Its target is *organogenesis* itself:

> The operative question can be posed as follows: In response to what forces and tensions, and at what threshold, will an ensemble of cells, defined by variable relations of adhesion or disconnection, fold into a particular morphological form and acquire particular cellular properties? Regenerative medicine works through the continuous variation of force fields and it is from this level up that it attempts to determine the emergence of particular tissue qualities (density, compressibility, elasticity), properties and forms (cell morphology and differentiation, organmorphology and structure). These forces might be biochemical, hydrodynamic or mechanical in nature.
>
> (Cooper 2008: 112)

In the process of explaining the transformation of molecular biology, Cooper enables us to ask what happens to security processes which take life as their referent object when life comes to be understood and formed by processes which replace, 'the technics of reproduction and substitution with an art of continuous modulation, in which form is plunged back into process, becoming continuously remorphable' (Cooper 2008: 115).

Security practices informed by understandings of living matter as a function of force fields of morphogenesis are not prophylactic practices. Regeneration rather than protection is more their condition of operation. Transformation rather than preservation is their object. They become preoccupied instead with securing (which now means engendering and manipulating) the contingent correlation of the events through which life forms morph and change. If Foucault's original exploration of biopolitics' *dispositifs de sécurité* identified the very eventalness of species existence as one of the base analytical categories of its biopoliticised security practices, molecularity has transformed eventalness as well. Its recombinant biopolitics is preoccupied with what we might call the securing of living material in its intrinsic 'pluripotence' as 'transformation event' (Lezaun 2006). The biopolitical securing of pluripotent life is thus concerned less with subjecting it to safety measures than with commanding its infinitely regenerable design. As the author of one recent overview put it:

> The fundamental principle of reparative medicine which governs our efforts to regenerate differentiated tissue is to organise the reparative circumstances to recapitulate the selected aspects of embryonic developmental sequence, including attempts to mimic the embryonic microenvironment in which tissue initiation, formation and expansion take place.
>
> (Arnold Caplan cited in Cooper 2008: 121)

As scientific historians at the Max Planck Institute continually remind us, molecular re-engineering is not simply aimed at the modification of metabolic processes but at their reprogramming: 'For the first time, it is on the level of instruction that metabolic processes are becoming susceptible to manipulation. Until that point was reached, medical intervention, even in its most intrusive physical, chemical and pharmacological forms, was restricted to the level of metabolic performance' (Rheinberger 2000: 25). What regenerative medicine wants to elicit, Cooper concludes, 'is the generative moment from which all possible forms can be re-generated … the moment of emergence in which all forms are virtually infolded' (Cooper 2008: 127). The referent object of biopolitical security strategies in the molecular age increasingly revolve around a new referent object of security, *the moment of emergence considered independent of its actualisation* (Cooper 2008: 127).

When life is reduced to code and the code has been mastered then the recruitment of life into its technoscientific manipulation and biopolitical strategisation shifts from the manipulation of forms of life into the command of regeneration itself. For these reasons we seem to be entering a new neo-baroque biopolitical order of the real in relation to the experience of the real now afforded by a recombinant biopolitical consensus revolving around molecular structure and hetero-genetic transformation. Once understood in terms of its recombinatory mechanisms, life becomes a matter of continuous mobile recomposition. When its science becomes less a matter of representation than of experimentation and design so also do its recombinant biopolitical security practices (Dillon and Reid 2001). To the question, what happens to security processes which take life as their referent object when life goes molecular? I answer that a recombinant biopolitics of security emerges oriented towards becoming an experimental scientific art of design.

Molecularisation: the recombinant biopolitical economy of life and death

The biopolitical economy of security has therefore undergone profound change since Foucault first began his analysis of it. While the base analytical categories which he identified appear to remain the same – circulation, economy, eventalness and above all contingency – their operationalisation in the form of specific problematics, political rationalities and governmental technologies now ranges from the global to the molecular. I have space to offer only one additional reflection. That reflection is this. The pluripotent life of recombinant biopolitical economy of security pluralises and problematises death as much as it does life. Together with the compression of morbidity it is now also offering a novel problematic for that economy, one around which new political rationalities and governmental technologies from health policy to life and health insurance are already beginning to revolve (Lobo-Guerrero 2011, 2012). That problematic is the very interface of life and death: from the pre-birth pluripotency of stem cells to the pre-death ageing of senescence. Recall my primary conclusion which is in any event worth restating here. Biopolitics of security secure by instantiating a general economy of the contingent throughout all the processes of re-productive circulation which impinge upon species existence as a whole.

In an age transforming itself, through the alliance of morbidity's compression with the regenerative promises of molecular medicine, the very contingency calculus of the biopolitical economy of contingency which characterises *dispositifs de sécurité* of biopolitics is thus being reworked in the context of the changing correlation of life and death prompted by the molecular revolution in particular. Despite his differentiation of power over life from power over death, Foucault was well aware of the correlation of life and death. In the celebrated chapter on 'Right of Death and Power over Life' in *The History of Sexuality* Foucault makes a variety of careful observations about the sovereign right over death and its development in response to the advent of biopolitics. The first is that it was never 'an absolute privilege' (Foucault 1998: 135). The second and more directly important is that with the advent of biopolitics: 'There has been a parallel shift in the right of death, or at least a tendency to align itself with the exigencies of a life administering power and to define itself accordingly' (Foucault 1998: 136). Pre-birth, pre-death (senescence) and the post-death of the comatosed neither dead nor alive, is precisely that domain where this formidable new power of death accrued by biopolitics, 'now presents itself as the counterpart of a power that exerts a positive influence on life, that endeavours to administer, optimise, and multiply it, subjecting it to precise controls and regulations' (Foucault 1998: 137). In assigning itself the power of administering life in and through its radical contingency, the biopolitics of security now administers the threshold of life/death: including, in particular, the new formation of life and the prolonged forestalling of death (Franklin 2001).

Conclusion: the biopolitical freedom of security

A complex baroquely expressed pact was, therefore, struck between security, knowledge and freedom in the political analytic of finitude of the modern age. The modern political imaginary has been circumscribed by the complex permutations through which that pact has mutated. Hitherto, our understanding of the pact has largely been dominated by the political discourse of the subject, of possessive individualism, and the political bond said to be contracted between the political subject and the modern state. But there are many ways in which the modern understands and promotes freedom, not least because the genealogy of freedom upon which the modern relies is as rich in Epicurean as it is, for example, in Cartesian, Hobbesian or Kantian motifs. Thus, the natural philosophy of early modern Europe found in Lucretius' *De Rerum Natura*, for example, allows the association of an inexplicable and unpredictable or uncaused swerve in the flow of originary matter (the clinamen) with the possibility of free will (see Serres 2000). Foucault observed that modern accounts of freedom in the citizen and the subject were also conceived in terms of what might be called the developmental freedom of its biopolitics (Senok *et al.* 2005).

What follows for freedom similarly follows for security. Modern conceptions of security have never been confined to the discourse of a subject presumed to exist prior to the power/knowledge relations that constitute it as the subject that it is, however that subject is conceived and at whatever level, individual or collective, it is presumed to operate. As he teaches us about the heterogeneity and problematicity of discourses of modern power, so also Foucault teaches us about the heterogeneity and problematicity of modern accounts of security and freedom. This teaching requires us not simply to apply Foucault but to go beyond Foucault. Foucault not only extended the register of what such practices meant and did. In the process he also alerted us to how the changing experiences of the true and the real which characterised the diverse complex of heterogeneous modern power relations were also deeply implicated in the detailed regulation of life locally and globally.

Foucault identified a different correlation of security and freedom than that which characterises the West's traditional politics of security, founded as it is on an allied politics of the subject.[9] There could be no more central question for politics, therefore, than that of the relation of freedom and security. While the very *mythos* of the modern state is founded in its claim to be a security provider, and the modern's commitment to life is founded in the promotion of life, their respective security mechanisms seem nonetheless fated to threaten the very ideal of political order, and the very valuation of life, which they are supposed to advance. What is well understood now in relation to the geopolitics of state security is evident also in biopolitics of security. Consistently hailed as part of the solution to the problematics of politics, freedom and justice, our baroque technologies of security have consistently proved themselves to be part of the problem instead. This applies as much to their biopolitical as to their geopolitical variants.

In order to promote life, biopolitical security practices must of course come to know life. But they do not simply insist on rendering life transparent to certain forms of knowing: such life, in whatever way resistant to being known, begins to pose a security problem to biopolitics. Their purpose is in addition of course to weigh life – to conduct a continuous assay of life – in order to determine which life is capable of regulating itself in the cause of its self-improvement, adaptation and change; and which forms of life have most to teach about these processes offering a kind of best practice of living and preferred forms of life. Different life forms display differing capacities in this respect. It therefore follows that some forms of life may be less capable or incapable, and even hostile or resistant, to self-regulation in the cause of self-improvement and adaptation. All life in some degree or another may have to be coached in its biopolitical self-governance and some life may have to be subject to more than coaching. Recalcitrant and intransigent forms of life may require punishment and correction. Ultimately, some life forms may be regarded as inimical to life itself and these will have to be eliminated. Foucault explores this logic and illustrates its operation in the account that he gives of race war (Foucault 2003). Others have illustrated it by reference to poverty programmes (Cruikshank 1999). 'Making life live' is as violent in its way as power over death.

The problematic of modern politics is, therefore, subtly misconceived when freedom and security are construed exclusively as contending values locked in a zero-sum game in which the one is advanced or lost at the expense of the other. Security is not opposed to freedom when freedom is understood as the contingent developmental freedom of biological existence. In its biological contingency, security does not possibilise freedom and freedom is not designed to constrain security. Here, instead, security becomes a set of mechanisms self-governing the very contingent properties of the freedom which biological entities are said to display; aspiring to autopoietic and complex adaptive systemic forms of freedom generically implicated in varying accounts of becoming, and of the continuous individuation of form. In the process, freedom becomes the principal mechanism by which biopolitical security is secured. Free to make life live freely, biopolitics of security will and do enforce that project violently. Through the regulation and enactment of the very aleatory character of species existence, modern freedom is the mechanism by which security is forcefully biopoliticised in the name of promoting the life and potentiality of the species. It does so, according to Foucault, to the point of its self-immolation. In Volume 1 of *The History of Sexuality*, he observes how a certain threshold of modernity is reached when it comes to wager the life of the species on its own (bio-)political strategies. Only when the interrogation of security acknowledges that it must become an interrogation of the intimate operational correlation of security and freedom in the biopolitics of security will it begin to engage the aporia of security which threatens the very project of political modernity itself; that in seeking to secure the promotion of human being as species being, modern security practices threaten its very existence. Making life live is a lethal business. Paradoxically, freedom from it may be required if species life is in fact to outlive its grip.

Notes

1 This is something which Agamben's biopolitics, for example, specifically does not do. Given its philosophical, juridical and, in particular, its messianic commitments, it cannot do so. See, for example, Thurschwell (2004); Ojakangas (2005); and Dillon (2005, 2007b).

2 Marketing strategies do of course aim to develop customer loyalty and identification with products, brands and stores. Indeed cards like Nectar cards are commonly called 'loyalty cards'. There is an extensive literature on brand identity. Again, I acknowledge the manifold practices of power relations continuously leaching into one another when analysing this particular operation of biopower.

3 Foucault expresses his hostility to species thinking and allied philosophies of becoming in his 'Theatrum Philosophicum' (1977: 165–96).

4 It is surprising how conventionally Foucault tells this story. Contrast it for example with historical sociologists like Giddens (1987) and Luhmann (1993: 236). Contrast it also with that told by political theorists like Tilly (1985: 85) and anthropologists like Asad (2003). Political theologians are especially hostile to the received account. See for example, Cavanaugh (2002) and Milbank (1990). The political theology of the modern state is a neglected topic, one whose significance has grown considerably with the advent of the so-called war on terror. A notable exception is Fletcher (2004).

5 Typically, see the following: Alberts and Czerwinski (1997); Alberts, Gartska and Stein (1999); Arquilla and Ronfeld (1997); Campen, Dearth and Goodden (1996); Czerwinski (1998).

6 This point is more implicit than explicit in Foucault's analysis. He was intrigued by the Epicureans and the Stoics and, in particular, by Gilles Deleuze's philosophy of the event. See in addition, Beistegui (2004); Serres (2000).

7 Colin McInnes once made the acute point that a biopolitics of security must have to contend also with the microbiological. He was right. Without enframing his pioneering work within the biopolitical analytic of security he has nonetheless provided powerful evidence of its operationalisation in health, security and foreign policy. See McInnes (2005a, 2005b); Lee and McInnes (2003). So also has Elbe (2003, 2006).

8 A large question opens up here. It not only concerns the relation between what we might call an Epicurean account of freedom – whose mark is that of the clinamen or swerve in being and beings – and a biologised account of freedom whose mark is that of circulation. It also concerns the complex ways in which the politically 'progressive' embracing of the Epicurean meets the politically 'regressive' embracing of the biological in a confluence which greatly complicates late modern accounts of freedom, form and event. See O'Keefe (2005). See, in addition to Serres (2000), Prigogine and Stengers (1984).

9 Giorgio Agamben (2001) comes close to doing so as well: 'Nothing is therefore more important than a revision of the concept of security as the basic principle of state politics.'

Bibliography

Agamben, Giorgio (1998) *Homo Sacer: Sovereign Power and Bare Life* (Stanford, CA: Stanford University Press).

Agamben, Giorgio (2001) 'Security and Terror', *Theory & Event*, 5(4): 1–6.

Agar, Nicholas (2004) *Liberal Eugenics: In Defence of Human Enhancement* (Oxford: Blackwell).

Alberts, David and Czerwinski, Thomas (eds) (1997) *Complexity, Global Politics and National Security* (Washington, DC: National Defense University).

Alberts, David, Gartska, John J. and Stein, Frederick P. (1999) *Network-Centric Warfare: The Face of Battle in the Twenty-First Century* (Washington, DC: National Defense University).

Arquilla, John and Ronfeldt, David F. (1997) *In Athena's Camp: Preparing for Conflict in the Information Age* (Santa Monica, CA: RAND).

Arquilla, John and Ronfeldt, David F. (2000) *Swarming & the Future of Conflict* (Santa Monica, CA: RAND).

Asad, Talal (2003) *Formations of the Secular: Christianity, Islam, Modernity* (Stanford, CA: Stanford University Press).

Barry, Andrew, Osborne, Thomas and Rose, Nikolas S. (eds) (1996) *Foucault and Political Reason: Liberalism, Neo-Liberalism, and Rationalities of Government* (Chicago, IL: University of Chicago Press).

Bartelson, Jens (1995) *A Genealogy of Sovereignty* (Cambridge: Cambridge University Press).

Beistegui, Miguel de (2004) *Truth and Genesis: Philosophy as Differential Ontology* (Bloomington: Indiana University Press).

Biersteker, T. J. and Eckert, S. E. (2006) *Countering the Financing of Terrorism* (Abingdon: Routledge).

Burchell, Graham, Gordon, Colin and Miller, Peter (eds) (1991) *The Foucault Effect: Studies in Governmentality – with two lectures by and an interview with Michel Foucault* (Chicago, IL: University of Chicago Press).

Campbell, C.A. (1998) 'Overview of Insurance Industry Issues and the Role of the Actuary'. Lecture presented to the Society of Actuaries, Session 1GS. Maui I Spring Meeting (15–17 June). *The Record*, 24.

Campen, Alan D., Dearth, Douglas and Goodden, Thomas (eds) (1996) *Cyberwar: Security, Strategy, and Conflict in the Information Age* (Fairfax, VA: AFCEA International Press).

Casey, Edward S. (1997) *The Fate of Place: A Philosophical History* (London: University of California Press).

Cavanaugh, William T. (2002) *Theopolitical Imagination* (London and New York: T. & T. Clark).

Cooper, Melinda (2006a) 'Pre-empting Emergence: The Biological Turn in the War on Terror', *Theory, Culture & Society*, 23(4): 113–35.

Cooper, Melinda (2006b) 'Resuscitations: Stem Cells and the Crisis of Old Age', *Body & Society*, 12(1): 1–23.

Cooper, Melinda (2008) *Life as Surplus: Biotechnology and Capitalism in the Neoliberal Era* (Seattle, WA: University of Washington Press).

Cruikshank, Barbara (1999) *The Will to Empower: Democratic Citizens and Other Subjects* (Ithaca, NY: Cornell University Press).

Czerwinski, Thomas J. (1998) *Coping with the Bounds: Speculations on Nonlinearity in Military Affairs* (Washington, DC: National Defense University).

Daston, Lorraine (1988) 'The Theory and Practice of Risk', in Daston, *Classical Probability in the Enlightenment* (Princeton, NJ: Princeton University Press), pp. 112–87.

Dillon, Michael (1996) *Politics of Security: Towards a Political Philosophy of Continental Thought* (London and New York: Routledge).

Dillon, Michael (2004) 'The Security of Governance', in Wendy Larner and William Walters (eds), *Global Governmentality: Governing International Spaces* (London and New York: Routledge), pp. 76–94.

Dillon, Michael (2005) 'Cared to Death: The Biopoliticised Time of Your Life', *Foucault Studies*, 1(2): 37–46.

Dillon, Michael (2007a) 'Governing Terror: The State of Emergency of Biopolitical Emergence', *International Political Sociology*, 1(1): 7–28.

Dillon, Michael (2007b) 'Governing through Contingency: The Security of Biopolitical Governance', *Political Geography*, 26(1): 41–7.

Dillon, Michael and Reid, Julian (2001) 'Global Liberal Governance: Biopolitics, Security and War', *Millennium – Journal of International Studies*, 30(1): 41–66.

Dillon, Michael and Reid, Julian (2009) *The Liberal Way of War: Killing to Make Life* (Abingdon: Routledge).

Doyle, Richard (2003) *Wetwares: Experiments in Postvital Living* (Minneapolis: University of Minnesota Press).

Elbe, Stefan (2003) *Strategic Implications of HIV/AIDS* (Oxford: Oxford University Press).

Elbe, Stefan (2006) 'Health and Security', in Alan Collins (ed.), *Contemporary Security Studies* (Oxford: Oxford University Press), pp. 334–45.

Elden, Stuart (2001) *Mapping the Present: Heidegger, Foucault and the Project of a Spatial History* (London: Continuum).

Elden, Stuart (2006) 'Rethinking Governmentality', *Political Geography*, 26: 29–33.

Elden, Stuart (2007) 'Governmentality, Calculation, Territory', *Environment and Planning D: Society and Space*, 25(3): 562–80.

Ericson, Richard V., Doyle, Aaron and Barry, Dean (2003) *Insurance as Governance* (Toronto: University of Toronto Press).

Fletcher, Paul (2004) 'The Political Theology of the Empire to Come', *Cambridge Review of International Affairs*, 17(1): 49–61.

Foucault, Michel (1977) *Language, Counter-Memory, Practice: Selected Essays and Interviews*, ed. Donald F. Bouchard (Ithaca, NY: Cornell University Press).

Foucault, Michel (1988) *Politics, Philosophy, Culture: Interviews and Other Writings, 1977–1984*, ed. Lawrence D. Kritzman (London and New York: Routledge).

Foucault, Michel (1989) *The Order of Things: An Archaeology of the Human Sciences* (London and New York: Routledge).

Foucault, Michel (1994) 'Polemics, Politics, and Problematizations', in *The Essential Works of Michel Foucault, 1954–1984*, vol. 1, ed. Paul Rabinow (London: Penguin), pp. 111–20.

Foucault, Michel (1996) 'Problematics', in *Foucault Live (Interviews, 1961–1984)*, ed. Sylvère Lotringer (New York: Semiotext(e)), pp. 416–22.

Foucault, Michel (1998) *The History of Sexuality*, vol. 1: *The Will to Knowledge* (London: Penguin).

Foucault, Michel (2003) *Society Must Be Defended: Lectures at the Collège de France, 1975–76* (London: Picador).

Foucault, Michel (2004a) *Naissance de la Biopolitique: Cours au Collège de France, 1978–1979* (Paris: Gallimard/Seuil).

Foucault, Michel (2004b) *Sécurité, Territoire, Population: Cours au Collège de France, 1977–1978* (Paris: Gallimard/Seuil).

Foucault, Michel (2007) *Security, Territory, Population: Lectures at the Collège de France, 1977–78*, ed. Michel Senellart (Basingstoke: Palgrave Macmillan).

Foucault, Michel (2008) *The Birth of Biopolitics: Lectures at the Collège de France, 1978–1979*, ed. Michel Senellart, François Ewald and Alessandro Fontana (Basingstoke: Palgrave Macmillan).

Franklin, James (2001) *The Science of Conjecture: Evidence and Probability before Pascal* (Baltimore, MD: Johns Hopkins University Press).

Giddens, Anthony (1987) *Social Theory and Modern Sociology* (Cambridge: Polity Press).

Gordon, Colin (1991) Governmental Rationality: An Introduction, Graham Burchell, Colin Gordon and Peter Miller (eds), *The Foucault Effect: Studies in Governmentality – with two lectures by and an interview with Michel Foucault* (Chicago, IL: University of Chicago Press), pp. 1–52.

Greenberg, Maurice, Weschler, William F. and Wolonsky, Lee S. (2003) *Terrorist Financing: Independent Task Force Report* (Washington, DC: Brookings Institution).

Grotius, Hugo (1925) *De jure belli ac pacis libri tres* (Oxford: Clarendon Press).

Hardt, Michael (2004) *Multitude: War and Democracy in the Age of Empire* (New York: Penguin).

Hayles, Katherine (1999) *How We Became Posthuman: Virtual Bodies in Cybernetics, Literature, and Informatics* (Chicago, IL: University of Chicago Press).

Heidegger, Martin (1962) *Being and Time* (Oxford: Basil Blackwell).

Larner, Wendy and Walters, William (2004) *Global Governmentality: Governing International Spaces* (London and New York: Routledge).

Lee, Kelley and McInnes, Colin (2003) *Health, Foreign Policy and Security*. Working Paper No. 1 (London: Nuffield Trust).

Lezaun, Javier (2006) 'Creating a New Object of Government: Making Genetically Modified Organisms Traceable', *Social Studies of Science*, 36(4): 499–531.

Lobo-Guerrero, Luis (2011) *Insuring Security: Biopolitics, Security and Risk* (Abingdon: Routledge).

Lobo-Guerrero, Luis (2012) *Insuring War: Sovereignty, Security and Risk* (Abingdon: Routledge).

Luhmann, Niklas (1993) *Risk: A Sociological Theory* (New York: Walter de Gruyter).

Machiavelli, Niccolò (2005) *The Prince* (Oxford: Oxford University Press).

McInnes, Colin (2005a) *Health and Foreign Policy in the UK: The Experience since 1997* (London: Nuffield Trust).

McInnes, Colin (2005b) *Health, Security and the Risk Society* (London: Nuffield Trust).
McInnes, Colin and Lee, Kelley (2006) 'Health, Security and Foreign Policy', *Review of International Studies*, 32: 5–23.
Maravall, José Antonio (1986) *Culture of the Baroque: Analysis of a Historical Structure*, trans. Terry Cochran (Minneapolis, MN: University of Minnesota Press).
Migdal, Joel S. (2004) *Boundaries and Belonging: States and Societies in the Struggle to Shape Identities and Local Practices* (Cambridge: Cambridge University Press).
Milbank, John (1990) *Theology and Social Theory: Beyond Secular Reason* (Oxford: Blackwell).
Mills, Catherine (2007) 'Biopolitics, Liberal Eugenics and Nihilism', in Steven DeCaroli and Matthew Calarco (eds), *Giorgio Agamben: Sovereignty and Life* (Stanford, CA: Stanford University Press), pp. 180–202.
Ojakangas, Mika (2005) 'Impossible Dialogue on Bio-Power: Agamben and Foucault', *Foucault Studies*, 2: 5–28.
O'Keefe, Tim (2005) *Epicurus on Freedom* (Cambridge: Cambridge University Press).
O'Malley, Pat (2004) 'Insurance, Actuarialism and Thift', in O'Malley, *Risk, Uncertainty, and Government* (London and Portland, OR: GlassHouse), pp. 117–34.
Ó Tuathail, Gearóid (1996) *Critical Geographies* (London and New York: Routledge).
Perrow, Charles (1999) *Normal Accidents: Living with High-Risk Technologies* (Princeton, NJ: Princeton University Press).
Petty, William (1690) *Political Arithmetick* (London: Printed for Robert Clavel at the Peacock, and Hen. Mortlock at the Phoenix in St. Paul's Church-yard: London).
Poitras, Geoffrey (2000) *The Early History of Financial Economics, 1478–1776: From Commercial Arithmetic to Life Annuities and Joint Stocks* (Cheltenham: Edward Elgar).
Prigogine, I. and Stengers, Isabelle (1984) *Order Out of Chaos: Man's New Dialogue with Nature* (London: Heinemann).
Rheinberger, Hans-Jorg (2000) 'Beyond Nature and Culture: Modes of Reasoning in the Age of Molecular Biology and Medicine', in Margaret M. Lock, Allan Young and Alberto Cambrosio (eds), *Living and Working with the New Medical Technologies: Intersections of Inquiry* (Cambridge: Cambridge University Press), pp. 19–30.
Sassen, Saskia (1991) *The Global City: New York, London, Tokyo* (Princeton, NJ: Princeton University Press).
Senok, A., Wilson, P., Reid, M., Scoular, A., Craig, N., McConnachie, A., Fitzpatrick, B., and MacDonald, A. (2005) 'Can We Evaluate Population Screening Strategies in UK General Practice? A Pilot Randomised Controlled Trial Comparing Postal and Opportunistic Screening for Genital Chlamydial Infection', *Journal of Epidemiology and Community Health*, 59(3): 198–204.
Serres, Michel (2000) *The Birth of Physics* (Manchester: Clinamen Press).
Thacker, Eugene (2004) *Biomedia* (Minneapolis: University of Minnesota Press).
Thurschwell, Adam (2004) 'Specters of Nietzsche: Potential Futures for the Concept of the Political in Agamben and Derrida'. Available at SSRN 969055.
Tilly, Charles (1985) 'War Making and State Making as Organised Crime', in Peter B. Evans, Dietrich Rueschemeyer and Theda Skocpol (eds), *Bringing the State Back In* (Cambridge: Cambridge University Press), pp. 169–91.
Vattel, Emer de (1883) *The Law of Nations = or, Principles of the Law of Nature, Applied to the Conduct and Affairs of Nations and Sovereigns; from the French of Monsieur de Vattel; from the new edition*, ed. Joseph Chitty and Edward D. Ingraham (Philadelphia: T. & J. W. Johnson).
Waldby, Catherine (2006) 'Umbilical Cord Blood: From Social Gift to Venture Capital', *Biosocieties*, 1(1): 55–70.
WHO (2003) 'Global Health Security: Epidemic Alert and Response'. Consensus Document on Epidemiology, May (Geneva: World Health Organization).

3

GOVERNMENT, ECONOMY AND BIOPOLITICS

> Modernity now appears as the horizon of an historical process that has to remain a gamble.
>
> (De Man 1971: 96)

Introduction: government and economy

For those wishing to understand Foucault's account of the birth of biopolitics, the lecture series of that title, which he gave at the Collège de France in 1978–9, comes as something of a disappointment. It was the last of three related lecture series. The first of the series was *Society Must Be Defended* (Foucault 2003), given during the course of 1975–6. The second was *Security, Territory, Population* (Foucault 2007), given during the course of 1977–8. In these Foucault had broached the emergence of the problematic of modern politics and security in general, as well as of governance, biopower and biopolitics in particular. Foucault anticipated the disappointment he feared that *The Birth of Biopolitics* (Foucault 2008) might excite, beginning Lecture 8 by apologising for allowing himself to be diverted from the announced plan of the lectures. 'I would like to assure you,' he said, 'that in spite of everything I really did intend to talk about biopolitics, and then, things being what they are, I have ended up talking at length, and maybe for too long, about neo-liberalism and neo-liberalism in its German form' (Foucault 2008: 185).

What *The Birth of Biopolitics* provides is a sustained interrogation of the fundamental theoretical and institutional revisions of economy, in Germany in particular, that emerged during the course of the late 1930s and 1940s, of how the category of economy as such was reconceived, and of how the relation of state and economy was also reconfigured. As Foucault observed in relation to the influential Freiburg School of economic theory, especially: 'The Freiburg School did not just develop an economic theory, or even a doctrine. It completely rethought the relation between government, the economy and politics, the whole art of government. And with good reason: it had to grapple with a considerable historical phenomenon' (Foucault 2008: 95n). That phenomenon was of course the political problematic of how to re-establish government in Germany, and found a new German state, after the successive catastrophes of its three previous regimes: the Kaiser Reich, the Weimar Republic and the Third Reich. It had also to do that not only in the context of Germany's experience of the crisis of capitalism, represented by the hyperinflation of the 1920s and the Great Depression of the 1930s, but also in response to the theologico-political cultural crisis precipitated by the collapse of the Protestant synthesis of the Wilhelmine period, signalled especially in the work of jurists like Schmitt and Kelsen as well as by Christian theologians like Petersen, Buber and Barth, and influential Jewish thinkers like Rosenzweig and Benjamin. While Foucault ignores this allied background to the developments that he does interrogate, for much of the time he is also preoccupied with the following three general lines of argument.

The first involved elaborating how the crisis of capitalism was simultaneously also a 'crisis of governance': specifically, a crisis in the very formulation of the problematic of governance as such, and not merely of a class-based revision of the instruments of state and economy, although such developments were clearly also involved as well. Government and economy were linked from the outset of Foucault's analysis of modern power relations. The two are locked in an historically emergent mutually disclosive belonging together that itself seems to depend, for Foucault, on a wide variety of equally historical dividing practices. These included those that have become foundational modern distinctions between state and society, as well as state and economy. Leaving aside, for the moment, the role economy played in his magnum opus *The Order of Things* (1989), his particular interest in these lectures lay in how the German Freiburg School, for example, treated government and economy together. I emphasise here that this was not new for Foucault. He had already made clear in the previous year's lecture course (*Security, Territory, Population*), for example, how much economy and governance were indebted to one another in the very posing of the problematic of government from the sixteenth to the eighteenth centuries – indeed, in the differentiation of what became known as 'the economy'. In respect of biopolitics, I will argue, in addition, and beyond Foucault, that the reduction of life to its biological properties compounds this economising of existence, because it universalises the utility that follows from the reduction of Life to species life. Just as species life is nothing if not utility-determined life, so also is modern economics nothing if not the science of utility.

Second, Foucault was concerned with how the crisis of economy and governance led, in Germany – and I will argue more generally throughout the societies of the North Atlantic basin, and their global imperial outworks – to the introduction of an 'economic polity'. (I will distinguish later between political economy and economic polity.) It suffices to say for the moment that political economy presupposes what I want to call a mixed constitution of regional ontologies – specifically here that of 'the state', that of 'society', and that of 'the economy' – in which, in addition to other political considerations, political legitimacy came to be founded on how government of the state was judged in respect of allowing 'economy' and 'society' to operate according to their own principles, as well as in terms of how effectively the state governed overall in relation to a wide variety of finite goals: what Foucault referred to as 'specific finalities' (Foucault 2007: 99). In the transformation of government and economy that occurred during the first half of the twentieth century, then, in Germany and the European Economic Community in particular, politics and the state, as well as aspirations to political union more generally, came to be founded on the establishment and orderly running of successful market-based economies.

However spurious some may think the autonomy to be, political economy claimed that politics and economy were independent regional ontologies of existence. In the institution of economic polities, however, the social and the political become largely derivative – one might almost say an epiphenomenon – of the successful dynamics of market-based economies. For Foucault, this is less the outcome of class-formation and class-based interest than a dynamic of the more generic posing of the problematic of government, and its 'economic' as well as governmental resolution. Government, for Foucault, came to operate fundamentally through economy, as well as being judged essentially in terms of how it governed in relation to 'the economy'. What began, however, as a profoundly important differentiation between state, society and economy in liberal regimes of power – since, among other things, government of the liberal state in the early modern period was conceived in terms of what Foucault calls the 'royal' questions of power – increasingly became a conflation of the three as the utilitarian logic of biopolitics worked its way through the comprehensive governmentalising, and indeed biopoliticising, of liberal states and societies; by which I mean, government became increasingly preoccupied with the governance of life through the promotion of life understood as species existence, itself grouped into contingent cohorts of individuals for which the generic term is 'population'.

It goes without saying, except to say it never goes without saying in these matters, that this is not the only way in which the problematic of government, politics and freedom can be configured. Neither is the finitude of the 'specific finalities' to which Foucault so intuitively referred the only way in which the analytic of finitude governing political modernity may be configured politically. I will return to these core issues later. Contesting biopolitics of security in the twenty-first century is, therefore, not only a matter of contesting the reduction of life to species existence; in the process it also requires contesting the analytic of finitude and facticity, as well as of freedom, which operationalises this reduction of mortal life to what is evidently a deeply aporetic and increasingly self-destructive form of biopolitical existence locally and globally.

Here, a profoundly important difference arises with Foucault that entails a significant reappraisal of the operation of the analytic of finitude. Yes, in as much as modern biopolitics of security is an expression of power/knowledge deeply influenced by changing accounts of the biological properties of human beings and the character of species existence in general, biopolitics of security in the twenty-first century is deeply enmeshed also in the aporetic impasse of the Kantian analytic of finitude, of which, epistemically, they are very much a political expression. This has one important caveat, addressed again later. If the Kantian analytic of finitude required the anthropological figure of Man to sustain its project, the modern epistemological and empirical referent for contemporary biopolitics is less Man than Life, Life understood in the changing terms of species existence. Biopolitics is, therefore, less a humanisation deriving from the anthropological machine of Man, than a remorseless speciation deriving from the techno-instrumentalist machine of biopower/knowledge.

However, the Kantian analytic of finitude is not the only expression of finitude. Here we cannot avoid Heidegger's *auseinandersetzung* with Kant on precisely this point:

> Hegel's step from Kant to absolute idealism is the sole consistency of the development of Western philosophy. This development is *possible* and *necessary* through Kant because the problem of human *Dasein*, finitude, did not become a real problem for him and thus not a central problem of philosophy because Kant himself – as the second edition of the *Critique of Pure Reason* shows – encouraged the path of working his way out of an uncomprehended finitude to appeasing himself with infinity.
>
> (Heidegger in Stambaugh 1992: 3)

The Kantian analytic of finitude, as Foucault details in *The Order of Things*, nonetheless did become a foundational modern referent point of departure for securing both governance and knowing, in fact for securing governance through knowing. For modern epistemology is not only a security project founded upon the modern knowing subject (Man); the continuous institution of such a subject, individually and collectively, increasingly became the modern's political project once the metaphysical enlightenment of the 1780s insisted, via Kant, on making such a subject foundational to modernity's very political imaginary (Campbell and Dillon 1993).

That said, one has also to consider how much further the dissolution of the subject has gone in the last ten years, in response not to Foucauldean, or any other kind of historicising, philosophising or theorising, but in response, instead, to dramatic technoscientific developments, and not only those of the digital and molecular revolutions, but also those taking place in the neurosciences as well. Such that one acute, numerate and scientifically literate observer has been moved to note:

> Once, not so long ago, there was an absolute opposition of self and other: an 'I' identical to itself, wholly present as an autonomous, indivisible, interior psyche against an external, amorphous collectivity of third persons outside the skin. Now the I/me-unit is dissolving, the one who says or who writes 'I' is no longer a single integrated whole, but multiple: a shifting plurality of

> distributed I-parts, I-roles, I-functions, and I-presences. Now the 'I' bleeds outward into the collective, and the collective introjects, insinuates, and internalises itself within the me. What was privately interior and individual is invaded by the public, the historical, the social. What was the outside world of events enters (*and reveals itself as always having entered*) the individual soul in the mode of personal destiny.
>
> (Rotman 2000: 99; emphasis added)

Inspired as I am by Foucault, however, I depart from him in two significant respects.

The first is in a sense merely historical. It concerns the changing account of species life that has arisen since the confluence of the digital and molecular revolutions of the twentieth century, and their widespread adoption throughout the biopolitics of security that now widely characterises the societies of the North Atlantic basin: a biopoliticisation of security that expresses itself not simply through an infinite war against terror but, more generically, also through the hyperbolicisation of surveillance and security that now characterises the economy of its biopolitical governance *tout court*. Here, in a sense, I continue to extend Foucault's project – of investigating the changing historical character of the biopolitical rationalities and governing technologies of biopower – by addressing how these became both radically more intensive and extensive by the first decade of the twenty-first century. I argue that the biopolitics of security of the twenty-first century goes hyperbolic because in construing life as a continuously emergent entity it thereby also institutes the governance of life as a continuous emergency. This differs from both the Schmittean and Benjaminian accounts of emergency informing much contemporary debate (Schmitt 2007, 2008; Benjamin 2004a, 2004b).

The second is, however, theoretical and analytical. It goes to the heart of Foucault's project and, indeed, the very critical project of the modern. It concerns my departure from Foucault in arguing that the modern – its biopolitical rationalities and governing technologies especially – are themselves founded upon and expressive of the temporal enframing of factical finitude. Further, the thought of factical finitude, politically, increasingly began to furnish the generative field of formation for modern politics from Machiavelli onwards; in other words, before Kant's Copernican revolution and the Kantian analytic of finitude (Kahn 2004). This allowed Foucault to claim that this marked the threshold epistemically of a modern episteme whose imminent demise he anticipated in the dissolution of the anthropological figure of 'Man': a figure that had once served, however briefly and unsuccessfully, to hold it together.

The empirico-transcendental double does not, however, characterise Kantian modernity alone. In the form of an empirico-eschatological double it has characterised political modernity since the very beginning of attempts to found politics in a reality of its own – divorced from that of the promise of eternal life which distinguished both the Christian message and the pastoral power that the Church sought to exercise over the flock which constituted its ecumene of belief. As Jacob

Taubes put it: 'the ontology, even of redemption, brings in need and presupposes the body of Christ, a commonality' (Taubes 2004: 56).

An epochal shift of a different sort therefore takes place as the West's politics of security and governance progressively shift from the soteriological message of the Christian era to that of the modern. That move is, I argue, a move away from a field of political and governmental formation revolving around the redemptive Christian thought of the eternal. Denied its referent point in such a soteriological account of the eternal, the modern emerged as that field of political formation that sought to give concrete political form to factical finitude. In this, it was actively seeking to institute and align itself with an alternative temporal and historical referent point to that which structured the political rationalities and governing technologies of Church and Kingship. Thus, if the soteriological politics of the Christian era was a politics of the eternal, and of creatures *sub specie aeternitatis*, the politics of security of the modern era is, instead, a politics of an infinity of temporal finitude.

In reintroducing the transcendental, however, Kant took sides in a long-standing theologico-political dispute that went back in the Christian tradition to Eusebius and Augustine (O'Donovan 1999; O'Donovan and O'Donovan 2000). Unlike others, including most notably Augustine (Markus 1988), Kant sided with those daring to promise that the Christian message of the security of eternal peace was a realisable prospect, at least in some degree, here on earth (de Vries 2006; Pranger 2006). Thus did the political and philosophical enterprises of the seventeenth and eighteenth centuries help pave the ways for that immanentising of the eschaton, so characteristic of contemporary technoscience in general, and its biopolitics of security in particular.

Foucault's third concern was to explore how post-war arrangements in Germany after 1945 also established what he called new 'fields of adversity' comprised of many 'surfaces of friction' whose negotiation characterised the detailed working out of its post-war order. 'Crises of governance', 'fields of adversity' and 'surfaces of friction' are important but widely neglected Foucauldean terms of art. Significantly inflecting Foucault's account of the operation of modern relations of power in general, and of biopolitics in particular, they offer additional insight into the dynamics of contemporary power relations. In particular, they further emphasise how power relations materialise in their contingent insufficiencies for Foucault. Consisting in a wide variety of immanent conflicts and contests, they contribute to explaining how, and why, power relations consistently misfire, miscalculate and are traversed by all manner of 'resistances', pragmatic as well as principled. If power is a web for Foucault, it is also ripped and riven. If it is a field of formation, that formation is always deforming, and reforming. If it is a grid of intelligibility, that intelligibility is not only always being fought for – the work of veridiction, or making true as Foucault puts it – it is continuously also forcefully contested as much on the inside as from the outside. Permeable to a whole series of different processes, technologies of power,

> are not immobile: they are not rigid structures aiming to immobilise living processes by their very immobility. Technologies of power are endlessly

> modified by the action of numerous factors ... This means that it is entirely possible to arrive at overall effects, not by concerted confrontation, but also by local or lateral or diagonal attacks that bring into play the general economy of the whole.
>
> (Foucault 2007: 119n–120n)

In apologising for taking a whole lecture course on biopolitics to talk about transformations in the category of economy that took place during the course of the mid-twentieth century, however, Foucault unwittingly helped to obscure the essential character of the relationship that obtains between biopolitics and economy. The apology encourages the listener/reader to think that, in talking about the revision of economy, Foucault is not talking about the character of biopolitics. But he is. The problem is that he does not explain how and why. Having intuited and detailed how biopolitics contributed to extending economy into the very heart of the problematic of modern government and politics (the work of the previous year), in *The Birth of Biopolitics* lectures he abandons further detailed explanation of the essential character of the relationship between biopolitics and economy in favour of a detailed account of the response to the mid-twentieth-century crises of both government and capital, in Germany especially.

Intimating that he had digressed from his theme, in fact Foucault had not. But he had failed to build the bridges necessary to link his earlier historical research into the early modern problematisation of government, politics and economy with this his more contemporary work. The failure was as much theoretical as it was historical. The exploration of biopolitics begun in *Security, Territory, Population* ends at the beginning of the nineteenth century. Foucault does not follow what happened, thereafter, until he addresses the crisis of governance and economy with which the German *ordoliberalen* were concerned in the first half of the twentieth century, and with which American anarcho-capitalists, as he calls them, of the Chicago School were later to be concerned. As he admits in the lectures, he does not much relate this work to a further elaboration of his analytic of biopolitics.

Albeit that we are indebted to him for posing the foundational question that introduces the analytic of biopolitics – what happens to power relations when they take species existence as their referent object of power – neither does Foucault much reflect theoretically on the essential consonance that obtains between economy and species existence. This is so, even as economic existence as economy comes to be formulated from the eighteenth century onwards in terms of the science of utility. Neither does he answer how that utility, operating at the centre ultimately of accounts of both economic and species life, weds these two together into the tight formation that increasingly characterises the economic polities of global liberal capitalism, and their governance.

I am not going to fill that historical gap, here, since my preoccupation, inspired by Foucault, and, as it were, in conversation with Foucault rather than slavishly pursuing and applying Foucault, is with the biopolitics of security as they operate at the beginning of the twenty-first century.

When the influence of biological and evolutionary theory on late Victorian thought is considered, however and for example, it is almost wholly in terms of the conflict version of Social Darwinism, the struggle for survival and the elimination of the unfit. But there was another representative and even more durable and influential link, connecting biological and socio-political theory to the mainstay of English new liberal ideology at the turn of the nineteenth and twentieth centuries. Ordinarily attributed, in particular, to the work ofT. H. Green, the incorporation of collectivism, community and social reform within this revision of English liberalism is attributable more to biological and evolutionary currents of thought. Biology and evolution were taken to supply 'empirical' evidence vindicating the transition of the liberal ethic towards a certain social and economic collectivism, while corroborating biopolitical assumptions, historically and philosophically, that according to Foucault had informed liberalism's political rationalities and governing technologies since the seventeenth and eighteenth centuries (Freeden 1976; Dillon and Reid 2009).

My specific concern here, with the biopolitics of security of the twenty-first century, does, however, direct attention to what I see as an important theoretical lacuna in Foucault. Recalling that he had already made the point, cited in the epigraph to this chapter, that modern forms of power relations revolve around the properties of the things they seek to direct, Foucault's historical research went on to show how the biological properties of human beings, specifically in as much as they concern, or express, utility in particular, became the object of governance. Hence, the advent of biopolitics for Foucault, at least, if not for Agamben (Mills 2008; Dillon and Reid 2009).

In his historically based research into the early modern emergence of biopolitics, Foucault is, therefore, driven by a central question. What happens to government and politics – to power relations more generally – when they take the biological properties of human beings as their referent object of government and politics? It is in the nature of such a question to ramify because the biological properties of human beings have been as subject to changing scientific interpretation as the nature of species themselves has been in question (Mayr 1997, 1998; Monod 1997; Jacob 1989a, 1989b). How, then, were discourses of government and politics impacted by the power/knowledge of the 'life sciences' that, emerging in the nineteenth century, developed exponentially in the twentieth century? What happened, we may also ask, to a form of government and politics whose political rationalities and governing technologies revolved around the infinity of temporal finitude of finite Life, construed as species existence, when the very concretisation of Life, thus understood, was itself transformed by technoscientific accounts of what it is to be a finite living thing? How have the new 'life sciences' impacted throughout the increasingly securitised economic governance of liberal societies more generally? How, in addition to the many ways in which these latest developments radically transformed the military strategic doctrines, operational concepts, training and organisation of the liberal way of war, has the explicitly killing game of the more generic hyperbolicisation of security that now characterises liberal governance, changed locally and globally (Dillon and Reid 2009)? Moreover, what are the wider ethical as well as political and governmental implications and consequences? Such questions can be posed in addition to those relating to the hyperbolicisation of security and surveillance of populations locally and globally that now excite such political concern. We are now faced with radically reducing the compass of existence – its very horizon of possibilities as well as the very corporeal experience of lived embodiment – to the political rationalities and governing technologies of an infinity of temporal finitude concretised as species existence, and pervasively governed through security technologies. A species existence operationalised, moreover, now, by and through the increasingly digitalised and molecularised political rationalities and governing technologies of the biopolitics of security of the twenty-first century.

Before I proceed with addressing this theoretical lacuna – beginning with the crucial chapter 'Man and his Doubles', from *The Order of Things*, where Foucault describes how 'the analytic of finitude' defines the aporetic structure of the modern episteme – I need, first, to recapitulate, how, in *Security, Territory, Population*, Foucault describes the problematic of government that was newly posed from the sixteenth century onwards. How he also sees that problematic of government installing economy at the heart of the problem of government. And, finally, how, as his research progresses, he notes the ways in which the very category of economy shifts as the problematic of government itself evolves, during the course of the seventeenth and eighteenth centuries, towards its mounting biopoliticisation.

The key point to bear in mind in relation to the category of economy – as indeed in relation to all the key categories involved in this analysis including later, for example, contingency as well as freedom, economy and government – is that 'economy' is not a universal. Economy has a history. Throughout these lecture series, Foucault can be seen tracking the changing historical materialisation of the category of economy alongside the changing historical materialisation of the intimately allied and equally historical category of 'government'. Later – in relation to the advent of biopolitics, and specifically of population as the original empirical referent object of biopolitical governance – he does this, also, with the category of the aleatory or the contingent (Dillon 2007). From the perspective of this analysis, all three – government, economy and contingency – are equally, also, involved with the emergence of security as the principle of formation around which the early modern problematisation not only of government and economy but also of 'freedom' came to revolve; albeit security has to be understood in terms of complex apparatuses of positivities concerned to promote the facilitation of an ever changing finite entity called Life, whose primary characteristic is precisely its 'finitude'. Hence, not only the biopolitics of security, but also the biopolitical analytic of finitude that I discuss in the next chapter, characterises what I propose to call 'Life and its Doubles'.

Government, economy, freedom

> The game of freedom and security is at the very heart of this new governmental reason whose general characteristics I have tried to describe. The

> problems of what I shall call the economy of power peculiar to liberalism are internally sustained, as it were, by this interplay of freedom and security.
> (Foucault 2008: 65)

Foucault spends the opening lectures of *Security, Territory, Population* exploring the emergence of the early modern problematisation of government, an art that was less concerned with statecraft and the development of early modern ideas about sovereignty, which long eclipsed it, than it was with responding to the generic crisis of the government of the conduct of conduct that was induced by the progressive dissolution of the medieval world. He pursues this task by examining a whole variety of neglected treatises – 'an immense, as well as monotonous, literature' (Foucault 2007: 89) – 'that do not exactly present themselves as advice to the prince, nor yet as political science' (Foucault 2007: 88). Emerging during the course of the sixteenth century, and continuing on through the seventeenth into the eighteenth centuries also, these present themselves as addressing the problem of 'government' instead: 'governing' a household, a family, a religious order, a province, children and souls as well as a state. These practices of government were construed as multifarious because many kinds of people were involved in governing, and government also sought to govern many different kinds of things. In that sense, then, the literature opened up the prospect of many different kinds of government, defined according to the properties of the thing or things to be governed, in relation to which the prince governing his state was only one.

'It seems to me,' Foucault continued, 'that all these problems, both in their intensity and multiplicity, are typical of the sixteenth century, and, putting it very schematically, are at the point of intersection of two movements' (Foucault 2007: 88). The first of these concerned a topic that Foucault has been much criticised for neglecting (Stoler 1995): 'the process that, dismantling feudal structures, organizes and sets up the great territorial, administrative, and *colonial* states' (Foucault 2007: 88–9; emphasis added). The second is different, yet, he says, complexly related to the first. It was that which, 'with the Reformation and Counter-Reformation, questions how one wishes to be spiritually directed here on earth for one's salvation' (Foucault 2007: 89).

The foundations, as well as the co-relations, long established via continuous historical dispute, and often violent conflict, between Church and Kingship – representative of Augustine's two cities, the City of Man and the City of God – had in fact, however, long been subject to that radical reappraisal, transformation and change which were to fashion the modern age; an age that was imprinted, of course, by what preceded it in these disputes as elsewhere. As Louis Dupré carefully noted, '[o]nly when the early humanist notion of creativity came to form a combustive mixture with the negative conclusions of nominalist theology [towards the end of the middle ages] did it cause the cultural explosion that we refer to as modernity. Its impact shattered the organic unity of the western view of the real' (Dupré 1993: 3) – including most notably for us, here, the real that underpinned doctrines of government and political legitimacy (Dupré 1993: 3). In Foucault's terms this meant that God no longer governed the world 'in the pastoral sense. He reigns over the world in a sovereign manner through principles' (Foucault 2007: 235).

More to the point, neither, increasingly, did God 'secure' the world. Progressively, the human was left to do that. Less determined in its everyday pastoral governance, as much as its political organisation by the soteriological promise of eternal salvation, the finite subject became architect of political constitution and governmental order without reference to the security of its supposed foundation in God's immutable natural law. Traumatised, in particular by the confessional religious wars of the seventeenth century, so influential, according to its foremost socio-political historian, to the institution of the baroque (Maravall 1986), securing order was not only sought through subjection to the 'mortal God' of the newly emerging sovereignties of the seventeenth and eighteenth centuries, and their baroque artifices (Maravall 1986; Lambert 2008; Egginton 2010), but also through the wider technologies of discipline and governance that such sovereigns themselves endorsed as vital to the securing of domestic and international peace (Hunter 2006; Foucault 1979).

Towards the end of the middle ages, nominalist theology had effectively removed God from creation (Dupré 1993; Blumenberg 1985). The divine became relegated increasingly to a supernatural sphere, separate from nature, with which it retained no more than a mystified causal link. Whereas, previously, meaning had been established in the very act of creation by an infinitely wise providential God, it progressively fell to the human mind to interpret a cosmos the structure of which was increasingly less given to the finite human subject as intelligible, than posited by that subject. Instead of being integral to a cosmological infinity, a finite human subject emerged responsible for specifying its truth, and positing its meaning; the very foundation of which became less the mystery of an eternal God than the positivity of human finitude. As the real, onto-theologically, split into two separate spheres – that of the mind which contained all intellectual determinations and that of all other being, which received them – so the ground of being as well as meaning, of the real as such, itself shifted from the eternal to the finite. Which is not to say that infinity as a reference for human kind disappeared. Rather, the modern thereby set itself the task of retrieving the infinite for itself from the only vantage point that became admissible to it: that of the base facticity of human finitude. If eternity was a correlate of the divine, infinity was a correlate of finitude.

Suffice to say, for the moment, that the infinity that it sought to recuperate – and in respect of which it becomes self-enchanting – while sharing properties with the eternity that preceded it in the Christian age, is, nonetheless, significantly different from it as well. For some, like Schmitt, this was merely secularising theological categories (Schmitt 2007, 2008; Meier 1998). The finite and the infinite are, however, a couplet. The very limit that demarcates finite things – call it mortality for human beings – cannot fail to gesture beyond itself. If the Christian world maintained a fundamental division between the eternal and the mortal, while thereby posing the problem of how to articulate the two, the modern age brings the eternal into the temporal by immanentising the eschaton. The end of times, the end of things, the play of the last things becomes a recurrent baroque trope of modern politics not least in its most empiricist biopolitical guises. The more it flees from the transcendental into the positivism of aesthetic and technoscientific artifice, the more aware it becomes of the ways in which it is haunted by that excess which, always already at play within the operation of limits, continuously threatens the security of positivistically established limits. The more, then, it also seeks to incorporate the beyond of those limits as the very Manichaean promise of the limit itself – the security of security against the existential enemy, the suppression of irrational insurrection securing perpetual peace for all rational beings.

For students of international relations, these two processes were ultimately involved also, of course, in the development of an international system of states and of modern international law. Foucault gives a rather traditional account of this development, and its allied political rationality of *raison d'état*, later in these lectures (Foucault 2007: Lectures 10–12). For students of security, however, it is important to note how these complex historical developments also witnessed the progressive decline of the Church's soteriological politics of security, the secular governmental expression of whose ecumene of belief, in Foucauldean terms, was its pastoral form of government (Foucault 2007: Lectures 5–9). Derived from the redeeming political economy of salvation (Kantorowicz 1997; Moore 1987; Kaufman 1990), life and death were figured, there, in relation to the promise of eternal life. Once the promise of eternal life lost its force as a generative principle of formation for the Church's governmental power – beginning with what Foucault calls 'the insurrection of conduct' witnessed in the sixteenth century (Foucault 2007: 227) – the material conditions of life, understood as finite life, progressively gained governmental sway thereafter, especially during the course of the seventeenth century (Hunter 2006; Kahn 2004; Kahn, Saccamano and Coli 2006; Oestreich 1982). This was finite life as positive conditions of possibility and not simply the fear of death (for Hobbes, paradigmatically, at the hands of other men).

This is not to ignore Kantorowicz's important argument concerning *The King's Two Bodies* (1997) in which he traced the early idea of the perpetuity of the state ('the king never dies') back to the construct of the Church as a mystical body, which in turn could never die. The way in which the *corpus mysticum* of the Church developed into the idea of a body politic is quite extraordinary, and abounds with paradox, as indeed does the subsequent perpetuity of the state. The thought of the finite remains contoured by its counterpart, the infinite, in modern as in pre-modern politics. The issue is not that of a move from the one to the other as if in a mere binary reversal. The issue is the refiguration of the whole ensemble that follows from a prioritisation and refiguration of the one over the other: the onset and pursuit, as it were, of a different principle of differentiation in which calculations to infinity positivistically instrumentalise what had previously been divinised (Rotman 1993a, 1993b, 2000). It is the changing nature of the baroque coupling of the finite/infinite that matters, and explains my concern with the empirico-eschatological ways in which it expresses itself in political modernity, whose thought of finitude is equally permeated by its own eschatological desires, promises and expressions. In that respect it is as important, therefore, to ask after the cultic and theological potentials in the political rationalities and governing technologies of the modern, in the way that Walter Benjamin did for example in his cryptic remarks on 'Capitalism as Religion' (2004a), as it is to ask after the political potentialities of the *re-ligio* as such, that which binds, to be found throughout theology in general and political theology in particular, for the Christian and post-Christian tradition at least.

By way of general contrast, then, we can say that, whereas the Church's soteriological politics of security governed by the promise of salvation were premised upon a political theology of eternity, modern politics of security came to be premised upon an analytic of finitude that, despite its increasing emphasis on the facticity of finitude, appropriated much pre-modern onto-theology via Man, as well, notably in respect of 'subjectivity', 'reason' and indeed salvation/security (Derrida 1995). Hence, any critique of our modern politics of security has to address the analytic of finitude that has been so formative in its production – and that includes its geopolitics of security as much as its biopolitics of security, of the infinity of the state, for example, as well as the infinity of the finite human.

For, a politics of security based upon the fractured, as well as finitised, onto-theological synthesis that has governed the modern's analytic of finitude is quite different from that based upon the onto-theological synthesis which once governed the Church's analytic of eternity. I turn my attention to that task in the next chapter, and in particular to the ways in which the biological properties of the human, understood as species existence, became the referent point for furnishing the finite properties, the empiricist default, to which modern politics of security looks as much for its political rationalities as its governing technologies. A default, I also want to say, that always already encodes its own peculiar eschatological desires and preoccupations as well. Perhaps the very best expression of this was given by the former US Secretary of Defense Donald Rumsfeld in his infamous remarks on the known and the unknown in which he concluded that it is the unknown unknowns that now preoccupy politics of security in the twenty-first century; a more baroque observation than that of Rumsfeld's unknown unknowns could hardly be imagined (Dillon and Reid 2009). A remark that finds its empirical expression, also, in the increasing obsession of the security communities of the North Atlantic basin with the catastrophic event, an event defined as catastrophic by virtue of the degree to which it is unforeseeable – exceeding the horizon of secure knowing, the beyond of normal forecasting, presaging the end of time as we know it – as well as materially devastating.

More generally stated, therefore, the problem that Foucault reconstituted historically from this literature was that of 'how to be governed, by whom, to what extent, to what ends, and by what methods' (Foucault 2007: 89). Contrasting this literature with that of Machiavelli's *The Prince*, for example, Foucault observes that while much of it was anti-Machiavellian, it was also a 'positive genre, with its specific objects, concepts and strategy' (Foucault 2007: 91). Since the literature that Foucault recovered ranged far and wide across the entire terrain of what was to be governed, and how, when it came to address the government of the state it also cast that problem within this wider frame of reference. Some of it offered a kind of typology of regional ontologies of government, each characterised by the distinguishing properties of the relevant area to be governed, 'each of them falling under a science or particular form of reflection' (Foucault 2007: 93). Thus the government of oneself was said to fall under the realm of morality while, in a tradition that goes back to the Greek *oikos*, the art of governing the family was a matter of economy, and that of governing the state a matter of politics (Foucault 2007: 93–4).

In that sense politics was distinguished as not entirely to do with morality, and not entirely to do with economy. And yet all three were correlated, since each was combined in the state. Since all these different modes of government arose within the context of society and the state, so there was the related problem of their relationship in their very diversity: 'It is within the state that the father governs his family, the superior governs the convent, and so on' (Foucault 2007: 93). 'There is then,' Foucault observed, 'both a plurality of forms of government and the immanence of practices of government to the state, a multiplicity and immanence of this activity that radically distinguishes it from the transcendent singularity of Machiavelli's Prince' (Foucault 2007: 93).

It was, therefore, understood that there was continuity upwards and downwards within the state between different areas subject to government: an 'upward continuity in the sense that whoever wants to be able to govern the state must first know how to govern himself, and then, at another level, his family, his goods, his lands, after which he will succeed in governing the state' (Foucault 2007: 94). Similarly, a downward continuity was posited in the sense that, 'when a state is governed well, fathers will know how to govern their families, their wealth, their goods and their property well, and individuals will also conduct themselves properly' (Foucault 2007: 94). It is, here, in particular, that the art of what became widely known and pursued in continental Europe as 'police' or *polizeiwissenschaft* emerged (Neocleous 2000). Police in this sense does not refer to the constabulary, or 'arm of the law', to which it almost exclusively refers in the Anglo-Saxon world. Foucault explains: 'This descending line, which means that the good government of the state affects individual conduct or family management, is what begins to be called "police"' (Foucault 2007: 94).

Pivotal in this emerging preoccupation with the general arts of government was the 'family', and central to the 'family' was 'economy' (Foucault 2007: 94). In sum,

> The art of government essentially appears in this literature as having to answer the question of how to introduce economy – that is to say the proper way of managing individual goods, and wealth, like the management of a family by a father who knows how to direct his wife, his children, and his servants, who knows how to make his family's fortunes prosper, and how to arrange suitable alliances for it – how to introduce this meticulous attention, this type of relationship between the father and the family, into the management of the state?
> (Foucault 2007: 94–5)

It is through the pivotal role that the family played in the science of 'police', then, and through which the science of 'police' was to play for a long time in the art of governing the state well, that economy became situated at the heart of the problematic of government as such. That is why, when Foucault repeats, '[t]he essential issue of government will be the introduction of economy into political practice' (Foucault 2007: 95), we should not assume that he is referring to either government or economy as universal categories: quite the contrary. He is referring to historically specific problematisations of both, extending into the science, especially, of 'police', ones that emerged as the ontopolitical assumptions, political rationalities, institutional forms and governing technologies of Europe's medieval order gave way to what came to be periodised as 'the modern'.

Not, as yet, however, the art of modern biopolitical government. Merely government. Several other developments had to take place before government was to revolve around the biological properties of the human way of being, and 'economy' was to escape the confines of 'family'. Foucault proceeded to explain how in the same set of lectures (Foucault 2007). First, it was the very semantic register of economy, as well as the technological apparatuses associated with the burgeoning commercial, and later early industrial, capitalism, that was to undergo significant transformation and change. The shift of semantic register also entailed a shift in the problematisation of government. Here, just as the art of exercising power in the form of economy comes to have 'economy' as its principal object, the word itself begins to signify something else: 'The word "economy"', Foucault says, 'designated a form of government in the sixteenth century' (Foucault 2007: 95). But the development of that art of government was constrained, on the one hand by the preoccupation with the massive abstract onto-theological problematic of sovereignty – together with its concentration on the building juridically and bureaucratically of modern state structures – and, on the other, by the restriction of economy to the narrow ensemble of the family which was 'too narrow, weak and insubstantial' to sustain further development of the idea and practices of 'economy', or, indeed, of the idea and practices of 'government' either.

However, according to Foucault, 'through a series of complex processes that are absolutely crucial for this genealogy', by the eighteenth century the term economy came to 'designate a level of reality and a field of intervention for government' (Foucault 2007: 95). The distinction being made, here, is crucial and pithily put: 'So, there you have what is governing and being governed' (Foucault 2007: 95). Indeed, Foucault is so pithy the distinction might escape notice. It is therefore worth detailing it a little more.

In other words, what 'governs' in the early modern emergence of the problematic of government is 'economy', understood in terms of the economy of household management or the management of the family. 'Governing a family,' Foucault explains, 'is not fundamentally directed toward the aim of safeguarding the family property, but essentially means having individuals who compose it, their wealth and prosperity, as their objective' (Foucault 2007: 97). It also means 'taking possible events, like deaths and births into account; it means considering the things one can do, such as alliances with other families' (Foucault 2007: 97). This was how the art of government, as the practice of the meticulous relations of family economy, was first conceived and applied. 'The essential main element, then, is this complex of men and things' (Foucault 2007: 97). Territory and property were of course involved, but they were not uniquely the referent objects of government conceived and practised in this way. The ensemble of men and things was the essential element: 'the territory and property being only variables' (Foucault 2007: 97). Moreover, law also 'recedes; or rather, law is certainly not the major instrument in the perspective of what government should be' (Foucault 2007: 99). 'Here again,' Foucault notes, 'we find the theme that recurs throughout the seventeenth century and is quite explicit in the eighteenth century texts of the *économistes* and physiocrats, which explain that the ends of government cannot be effectively achieved by means of the law' (Foucault 2007: 99).

By the eighteenth century, economy thus becomes posited as something that did not so much distinguish the reality of the family, however, but was, itself, constituted as an independent lived reality (a regional ontology in its own right). As such, it was said to be subject to its own autonomous principles of formation, laws and dynamics, the better governance of which became posed as an issue for the state. The answer to whose better government, moreover, was to become summarised in the doctrine of laissez-faire. Note the shift. In the one context, that of early modern government, economy is what governs. In the other context, what became known as classical political economy, 'the economy' is what is to be governed, and governed best in accordance with its own posited reality. By the twenty-first century, however, it is now the economy that largely governs.

As government became linked to the developing administrative apparatuses of the newly emergent centralised monarchies of the seventeenth and eighteenth centuries, so also did it begin to draw upon an allied form of knowledge, pursued assiduously in England as much as in continental Europe (Daston 1988; Hacking 1975, 1990) as well as in the colonies of North America and, later, the newly independent republic of the United States (Houston 2003; Cassedy 1969). This was originally known as political arithmetic. Once a mechanism for sovereigns and colonisers it was soon to become the science of statistics, available to progressives and reformers as well (Buck 1977, 1982; McLean 2004; Petty 1997; Phiddian 1996; McCormick 2007). Although not exclusively, such expertise was associated also in its earliest days with mercantilism. Foucault tells us that the first attempt at providing a political rationality of state power in terms of this developing account of government was mercantilism. Mercantilism subsequently fails, he says, 'precisely because it took the sovereign's might as its essential objective' (Foucault 2007: 102). And its instruments were those of the sovereign, 'law, edicts and regulations, that is to say the traditional weapons of sovereignty' (Foucault 2007: 102). It was as if, Foucault is saying, mercantilism could not spin the arts of government any further because its horizon of thought, as well as its practice, was constrained by its focus on increasing 'the might of the sovereign', rather than developing the powers of government.

In a paradox that is equally evident today, focusing on the concerns of sovereign power has the effect of limiting one's utility to sovereignty. For, despite its preoccupation with the metaphysics of power and with the policies of territorial sovereignty, and as Foucault also argues in these lectures (Foucault 2007: 109), modern sovereignty is so thoroughly governmentalised – preoccupied with all the modes of regulation required to govern the ensemble of men and things, perhaps most when it proclaims its limited intentions in respect of them – that it has an insatiable appetite for precisely that inventiveness in the development of power relations to be derived from maintaining critical distance from sovereignty. From the very earliest days of modern sovereignty, it was apparent that sovereigns could not achieve their modernising ambitions – not least those concerning the very institutionalisation of modern state sovereignty itself – without that development of power relations derived from outwith the onto-theological and institutional horizons of sovereignty (Foucault 1979). If you are in the market for strategic ideas concerning power relations, be sure the sovereign will sniff you out. But mercantilism tried 'to introduce the possibilities given by a reflected art of government within an institutional and mental structure of sovereignty that blocked it' (Foucault 2007: 102). In respect of economy, as well, mercantilism was confined within the horizons of an economics, 'which at that time only ever referred to the management of a small *ensemble* comprising the family and the household' (Foucault 2007: 103).

Biopolitics therefore arises, in Foucault's account, as the movement by which the constraints on the early modern arts of government and economy were released by the shift to the biological properties of human being, in general, and to the early empirical object of biopolitical governance – population – in particular. A population is a multiplicity, but it is not a community. Rather, a population is a mere cohort of individuals that, *en masse*, displays observable properties and characteristic features. It was these that became the new and more fruitful object of the arts of both government and economy, because they became the object of an emerging epistemic concern with the characteristics of species existence and behaviour. Indeed, since economy was already installed at the heart of government, what Foucault goes on to describe was a dual process: the development of the economy of government *parri passu* with the government of economy.

By the mid-eighteenth century, especially in England and its North American colonies, instead of opposing the instruments of political arithmetic (censuses and statistics) as instruments of the tyranny of sovereign power, emancipators and reformers, religious and political free thinkers alike, embraced statistics as a means of advancing their progressive and emancipatory economic as well as political ambitions (Cassedy 1969; Rusnock 2002). This occurred in respect, on the one hand, of lobbying for constitutional political order to reflect new wealth-based classes and interests, as well as, on the other hand, of lobbying for government to address newly emerging economic and social problems (Buck 1977, 1982).

How then was the art of government released from its constraints, asks Foucault? His general answer was through a variety of factors including: 'demographic expansion of the eighteenth century, which was linked to the abundance of money, which was itself linked to the abundance of agricultural production' (Foucault 2007: 103).

But there was, of course, more besides. His is only an indicative sketch, by no means a rigorous or exhaustive history. Had it been so he might well have pursued his observation about the increase of money since that too, once again in the British empire especially, was a function of public credit (Dickson 1967; Poitras 2000; Brantlinger 1996) and insurance (Clark 1999; Hendriks 1851; Lobo-Guerrero 2011, 2012). The transformation of economy was, in short, no simple matter of the revolution in the means of production. Neither, of course, were its effects confined to economy alone. Foucault was tracking a transformation in the correlated problematisation of both government and economy. In particular the revolution in economy also entailed the invention of new forms of money, and the greater access to credit, especially public credit, which was itself so closely associated not only with imperial expansion and wars for global hegemony, between England and France, but also with the correlate rise of the 'nation' state and its arts of government (Brewer 1988).

Then, as now, the significance of public credit was widely acknowledged: for example, by David Hume among many others (Laursen and Coolidge 1994). Then, as now, great scepticism was also expressed about the alchemical powers of money, debt and credit. Defoe's ambivalence, expressed at the beginning of the eighteenth century, was shared by most literati and conservatives at the time:

> As Light vanishes into Darkness; [Credit] comes with Surprize, and goes without Notice ... while it dwells in a Nation, it doubles their strength, they can fit out Fleets without Money, have Money without Funds, and Funds without Deficiencies. What has this invisible Phantom done for this Nation, and what miserable Doings were there here before without her?
>
> (Defoe in Brantlinger 1996: 1)

Then, as now, notwithstanding the many differences one can draw between the global imperial campaigns of the eighteenth and twenty-first centuries, war as the traditional occupation of sovereignty could not be conducted without the money that the invention of public credit and the development of all manner of new financial instruments introduced throughout this period (Dickson 1967; Brewer 1988; Van Der Heijden 2006; Ferguson 2008).

The development of an adequate somewhat stabilising or at least stable theorisation of national debt and public credit, for example, came 'only after the bursting of the South Sea Bubble in 1721, and really only after the middle of that century, in the work of David Hume, Adam Smith, Henry Thornton and others' (Brantlinger 1996: 7; Brewer 1988; Dickson 1967). Marx, too, of course, recognised that the capitalist mode of production is based on public credit, or on the modern money form. As an expression of the wider uncanny significatory character of money in the new games of risk and chance, introduced throughout the eighteenth century (Simmel 1991; Rotman 1993b; Shell 1982; Ingham 2004), public finance in particular (public debt) became intimately allied also to political legitimacy, to national indebtedness, and to what Hume called 'opinion'. As Brantlinger among other analysts observed: 'The warring nation states of Europe, energized by expansion abroad and industrialization at home, were able to carry on their war making, imperializing, and industrializing only on the basis of pervasive, new fiscal arrangements' (Brantlinger 1996: 22). And while it remains to be documented in detail, the same can be said with equal force about the phenomenon of global debt financing and the imperial war making conducted by the United States and Great Britain, in particular, during the course of the first decade of the twenty-first century (Ferguson 2009; Martin 2007). Biopolitics articulate with geopolitics through a wide variety of mechanisms, not least, as in this instance, the uncanny games of chance associated with money and war, contingency and risk.

That noted, Foucault was especially concerned with the way in which the problematic of population displaced that of family to become the pivot also around which government, economy and statistics were to change as well. This is the crucial passage. It requires quoting at length:

> This being the general framework, I think we can say more precisely that the unlocking of the art of government was linked to the emergence of the problem of population. Or, let's say that there is a quite subtle process, which we should try to reconstruct in detail, in which we can see how the science of government, the re-focusing of the economy on something other than the family, and the problem of population were all connected. It is through the development of the science of government that the economy could be re-focused on a level of reality that we now describe as the economic; and it is again through the science of government that the specific problem of population could be identified. But we could also say that it is thanks to the perception of the specific problems of the population, and thanks to the isolation of the level of reality that we call the economy, that it was possible to think, reflect, and calculate the problem of government outside the juridical framework of sovereignty.
>
> (Foucault 2007: 103–4)

Similarly, as it refocused increasingly on the quantifiable regularities displayed by population as such – 'its death rate, its incidence of disease, its regularities of accidents (Foucault 2007: 104) – so also did statistics detach itself not only from the family, but also from its earlier identification with the political arithmetic of sovereigns (Buck 1977, 1982):

> Statistics also shows that the population also involves specific aggregate effects and that these phenomena are irreducible to those of the family: major epidemics, endemic expansions, the spiral of labour and wealth. Statistics [further] shows that, through its movements, its customs and its activity, population has specific economic effects ... Apart from some residual themes ... the family disappears as the model of government.
>
> (Foucault 2007: 104)

What begins to emerge, instead, his analysis shows, is not only the burgeoning of population as a novel empirical referent stimulating both the arts of government as well as the sciences of economy and statistics – 'The constitution of a knowledge (*savoir*) of government is absolutely inseparable from the constitution of knowledge of all the processes revolving around population in the wider sense of what we now call political economy' (Foucault 2007: 106) – but also the emergence and further development ultimately of the wider problematic of governing by reference to the biological properties of species existence in general: biopower and biopolitics.

Foucault is, however, also careful to note that 'sovereignty is absolutely not eliminated by the emergence of a new government that has crossed the threshold of political science. The problem of sovereignty is not eliminated, on the contrary, it is made more acute than ever' (Foucault 2007: 107). And one of the ways in which it is made more acute is precisely in terms of its relation to these new burgeoning problematisations and arts of government. So, he emphasises, 'we should not see things as the replacement of a society of sovereignty by a society of discipline, and then of a society of discipline by a society, say, of government. In fact we have a triangle: sovereignty, discipline, and governmental management, which has population as its main target and apparatuses of security as its essential mechanism' (Foucault 2007: 107–8).

Coursing throughout all three, materially contributing towards their future development, was increasing reliance upon the correlate thought of factical finitude. But the thought of finitude did not prove to be a fixed and determinate body of thought. Neither does finitude itself prove to be as secure and reliable a reference point for Life and its governance as it might appear, and as some would claim.

Bibliography

Benjamin, Walter (2004a) 'Capitalism as Religion', in Benjamin, *Selected Writings*, ed. Marcus Paul Bullock and Michael William Jennings (Cambridge, MA: Belknap Press), pp. 288–91.

Benjamin, Walter (2004b) 'Critique of Violence', in Benjamin, *Selected Writings*, ed. Marcus Paul Bullock and Michael William Jennings (Cambridge, MA: Belknap Press), pp. 236–52.

Blumenberg, Hans (1985) *The Legitimacy of the Modern Age* (Cambridge, MA: MIT Press.).

Brantlinger, Patrick (1996) *Fictions of State: Culture and Credit in Britain, 1694–1994* (Ithaca, NY: Cornell University Press).

Brewer, John (1988) *The Sinews Of Power: War, Money, and the English State, 1688–1783* (Cambridge, MA: Harvard University Press).

Buck, Peter (1977) 'Seventeenth-Century Political Arithmetic: Civil Strife and Vital Statistics', *Isis*, 68(1): 67–84.

Buck, Peter (1982) 'People Who Counted: Political Arithmetic in the Eighteenth Century', *Isis*, 73(1): 28–45.

Campbell, David and Dillon, Michael (eds) (1993) *The Political Subject of Violence* (Manchester: Manchester University Press).

Cassedy, James H. (1969) *Demography in Early America: Beginnings of the Statistical Mind, 1600–1800* (Cambridge, MA: Harvard University Press).

Clark, Geoffrey Wilson (1999) *Betting on Lives: The Culture of Life Insurance in England, 1695–1775* (Manchester: Manchester University Press).

Daston, Lorraine (1988) *Classical Probability in the Enlightenment* (Princeton, NJ: Princeton University Press).

De Man, Paul (1971) *Blindness and Insight* (Minneapolis, MN: University of Minnesota Press).

Derrida, Jacques (1995) *On the Name*, ed. Thomas Dutoit (Stanford, CA: Stanford University Press).

Dickson, P. G. M. (1967) *The Financial Revolution in England: A Study in the Development of Public Credit, 1688–1756* (New York: St. Martin's Press).

Dillon, Michael (2007) 'Governing through Contingency: The Security of Biopolitical Governance', *Political Geography*, 26(1): 41–7.

Dillon, Michael and Reid, Julian (2009) *The Liberal Way of War: Killing to Make Life* (Abingdon: Routledge).

Dupré, Louis K. (1993) *Passage to Modernity: An Essay in the Hermeneutics of Nature and Culture* (New Haven: Yale University Press).

Egginton, William (2010) *The Theater of Truth: The Ideology of (Neo)Baroque Aesthetics* (Stanford, CA: Stanford University Press).

Ferguson, Niall (2008) *The Ascent of Money: A Financial History of the World* (New York: Penguin).

Foucault, Michel (1979) *Discipline and Punish: The Birth of the Prison* (Harmondsworth: Penguin).

Foucault, Michel (1989) *The Order of Things: An Archaeology of the Human Sciences* (London and New York: Routledge).

Foucault, Michel (1997) 'Society Must Be Defended', in *The Essential Works of Foucault, 1954–1984*, ed. Paul Rabinow, vol. 1 (New York: New Press), pp. 59–66.

Foucault, Michel (2003) *Society Must Be Defended: Lectures at the Collège de France, 1975–76* (London: Picador).

Foucault, Michel (2007) *Security, Territory, Population: Lectures at the Collège de France, 1977–78*, ed. Michel Senellart (Basingstoke: Palgrave Macmillan).

Foucault, Michel (2008) *The Birth of Biopolitics: Lectures at the Collège de France, 1978–1979*, ed. Michel Senellart, François Ewald and Alessandro Fontana (Basingstoke: Palgrave Macmillan).

Freeden, Michael (1976) 'Biological and Evolutionary Roots of the New Liberalism in England', *Political Theory*, 4(4): 471–490.

Hacking, Ian (1975) *The Emergence of Probability: A Philosophical Study of Early Ideas about Probability, Induction and Statistical Inference* (Cambridge: Cambridge University Press).

Hacking, Ian (1990) *The Taming of Chance* (Cambridge: Cambridge University Press).

Han, Béatrice (2002) *Foucault's Critical Project: Between the Transcendental and the Historical* (Stanford, CA: Stanford University Press).

Hendriks, Frederick (1851) Contributions to the History of Insurance and of the Theory of Life Contingencies, with a Restoration of the Grand Pensionary de Wit's Treatise on Life Annuities (London: printed by C. and E. Layton).

Houston, Alan (2003) *Population Politics: Benjamin Franklin and the Peopling of North America* (La Jolla, CA: Center for Comparative Immigration Studies, University of California, San Diego).

Hunter, Ian (2006) *Rival Enlightenments: Civil and Metaphysical Philosophy in Early Modern Europe* (Cambridge: Cambridge University Press).

Ingham, Geoffrey K. (2004) *The Nature of Money* (Cambridge: Polity Press).

Jacob, François (1989a) *The Logic of Life: A History of Heredity* (Harmondsworth: Penguin).

Jacob, François (1989b) *The Possible and the Actual* (Harmondsworth: Penguin).

Kahn, Victoria (2004) *Wayward Contracts: The Crisis of Political Obligation in England 1640–1674* (Princeton, NJ: Princeton University Press).

Kahn, Victoria Ann, Saccamano, Neil, and Coli, Daniela (eds) (2006) *Politics and the Passions: 1500–1850* (Princeton, NJ: Princeton University Press).

Kantorowicz, Ernst Hartwig (1997) *The King's Two Bodies: A Study in Mediaeval Political Theology* (Princeton, NJ: Princeton University Press).

Kaufman, Peter Iver (1990) *Redeeming Politics* (Princeton, NJ: Princeton University Press).

Lambert, Gregg (2008) *On the (New) Baroque* (Aurora, CO: Davies Group).

Laursen, John Christian and Coolidge, Greg (1994) 'David Hume and Public Credit: Crying Wolf?' *Hume Studies*, 20(1): 143–50.

Lobo-Guerrero, Luis (2011) *Insuring Security: Biopolitics, Security and Risk* (Abingdon: Routledge).

Lobo-Guerrero, Luis (2012) *Insuring War: Sovereignty, Security and Risk* (Abingdon: Routledge).

McCormick, Ted (2007) 'Transmutation, Inclusion, and Exclusion: Political Arithmetic from Charles II to William III', *The Journal of Historical Sociology*, 20(3): 259–78.

McLean, Stuart (2004) *The Event and its Terrors: Ireland, Famine, Modernity* (Stanford, CA: Stanford University Press).

Maravall, José Antonio (1986) *Culture of the Baroque: Analysis of a Historical Structure*, trans. Terry Cochran (Minneapolis: University of Minnesota Press).

Markus, R. A. (1988) *Saeculum: History and Society in the Theology of St Augustine* (Cambridge: Cambridge University Press).

Martin, Randy (2007) *An Empire of Indifference: American War and the Financial Logic of Risk Management* (Durham, NC: Duke University Press).

Meier, Heinrich (1998) *The Lesson of Carl Schmitt: Four Chapters on the Distinction between Political Theology and Political Philosophy* (Chicago, IL: University of Chicago Press).

Mills, Catherine (2008) 'Playing with Law: Agamben and Derrida on Postjuridical Justice', *South Atlantic Quarterly*, 107(1): 15–36.

Monod, Jacques (1997) *Chance and Necessity: An Essay on the Natural Philosophy of Modern Biology* (London: Penguin).

Moore, R. I. (1987) *The Formation of a Persecuting Society: Power and Deviance in Western Europe, 950–1250* (Oxford: Blackwell).

Neocleous, Mark (2000) *The Fabrication of Social Order: A Critical Theory of Police Power* (Sterling, VA: Pluto Press).

O'Donovan, Oliver (1999) *The Desire of the Nations: Rediscovering the Roots of Political Theology* (Cambridge: Cambridge University Press).

O'Donovan, Oliver and O'Donovan, Joan Lockwood (2000) *From Iranaeus to Grotius: A Source Book in Christian Political Thought* (Grand Rapids, MI: William B. Edermans).

Oestreich, Gerhard (1982) *Neostoicism and the Early Modern State* (Cambridge: Cambridge University Press).

Petty, William (1997) *The History of the Survey of Ireland, commonly called the Down Survey* (London and New York: Routledge).

Phiddian, Robert (1996) 'Political Arithmetick: Accounting for Irony in Swift's *A Modest Proposal*', *Accounting, Auditing & Accountability Journal*, 9(5): 71–83.

Poitras, Geoffrey (2000) *The Early History of Financial Economics, 1478–1776: From Commercial Arithmetic to Life Annuities and Joint Stocks* (Cheltenham: Edward Elgar).

Pranger, M. B. (2006) 'Politics and Finitude: The Temporal Status of Augustine's Civitas Permixta', in Hent de Vries and Lawrence Eugene Sullivan (eds), *Political Theologies: Public Religions in a Post-Secular World* (New York: Fordham University Press), pp. 113–21.

Rotman, Brian (1993a) *Ad infinitum … The Ghost in Turing's Machine: Taking God Out of Mathematics and Putting the Body Back In: An Essay in Corporeal Semiotics* (Stanford, CA: Stanford University Press).

Rotman, Brian (1993b) *Signifying Nothing: The Semiotics of Zero* (Stanford, CA: Stanford University Press).

Rotman, B. (2000) *Mathematics as Sign: Writing, Imagining, Counting* (Stanford, CA: Stanford University Press).

Rusnock, Andrea Alice (2002) *Vital Accounts: Quantifying Health and Population in Eighteenth-Century England and France* (Cambridge: Cambridge University Press).

Schmitt, Carl (2007) *The Concept of the Political* (Chicago, IL: University of Chicago Press).

Schmitt, Carl (2008) *Political Theology II: The Myth of the Closure of Any Political Theology* (Cambridge: Polity Press).

Shell, Marc (1982) *Money, Language, and Thought: Literary and Philosophic Economies from the Medieval to the Modern Era* (Baltimore: Johns Hopkins University Press).

Simmel, Georg (1991) *The Philosophy of Money* (London and New York: Routledge).

Stambaugh, Joan (1992) *The Finitude of Being* (Albany, NY: State University of New York Press).

Stoler, Ann Laura (1995) *Race and the Education of Desire: Foucault's History of Sexuality and the Colonial Order of Things* (Durham, NC: Duke University Press).

Taubes, Jacob (2004) *The Political Theology of Paul* (Stanford, CA: Stanford University Press).

Van Der Heijden, Manon (2006) 'State Formation and Urban Finances in Sixteenth and Seventeenth-Century Holland', *Journal of Urban History*, 32(3): 429–50.

Vries, Hent de (2006) 'Introduction', in Hent de Vries and Lawrence Eugene Sullivan (eds), *Political Theologies: Public Religions in a Post-Secular World* (New York: Fordham University Press), pp. 1–88.

4
UNDERWRITING SECURITY

> What if security unleashed your full potential? At Unisys were looking at security in an entirely new way. Security is no longer a defensive measure. It's an enabling catalyst for achievement. Unisys Secure Business operations help to unleash your full potential.[1]

This chapter continues with the exploration of how biopolitics of security takes factically finite Life as its referent object of security by detailing how it enframes 'risk' as a biopolitical security technology. Since contingency is constitutive of the life of emergent being, this chapter explains that a biopolitics of security that seeks 'to make life live' cannot secure life against contingency but must secure life through governmental technologies of contingency. Risk is one of these technologies. The chapter also explains how risk came to pervade the biopolitics of security of the twenty-first century, and how, through the way in which it is traded on the capital markets, it has begun to acquire the properties of money. The chapter closes by further describing how biopolitics of security differs from traditional prophylactic accounts of security, and how biopolitics of security exceeds the liberal political thinking that rationalises and legitimates them. In the process, many of the political and governmental artifices for 'making sense' mutate from baroque to neo-baroque.

Enframing risk biopolitically

The title of this chapter puns David Campbell's pioneering work *Writing Security* (1998). Without disputing the continuing salience of the politics of identity in the problematisation and operationalisation of contemporary security practices, without also dismissing the continuing fixation with the baroque politics and mythology of sovereign subjectivity, 'Underwriting security' seeks an additional shift in our analytical focus, and field of application, of biopolitics of security, to the ways in which the biopoliticisation of security installs risk as one of its single most important devices. Underwriting, a term taken from the insurance industry, captures the essence of how risk operates as an assemblage of mechanisms for measuring and commodifying exposure to contingency (Baker and Simon 2002; Clark 1999; Ericson and Doyle 2003; Ericson, Doyle and Dean, 2003; Ewald 1986; O'Malley 2006). As Luis Lobo-Guerrero has argued in detail, insurance has long been an integral part of the provision of biopolitical security (2011, 2012). Since, biopolitically speaking, contingency is constitutive of what it is to be a living thing, the referent object of biopolitics – life as species being – cannot be secured against contingency. Biopolitically, it is instead secured through contingency (Dillon 2007). Risk is one of the single most important devices by which this biopoliticised securitisation is currently pursued.

Since this biopoliticising of security is also a historical phenomenon, it is a project and not an accomplishment. It is therefore a plural and changing thing, and it changes according to different accounts of what it is to be a 'living thing'.

In his justly celebrated works, Nobel Prize winner François Jacob surveys the changing account of living beings. Exploring how life has been problematised from the beginning of the modern age through the eighteenth and nineteenth centuries up to the twentieth century, Jacob (1989) observes that the modern age begins by breaking with the centuries-old Aristotelian and Christian accounts of being and beings. He explains how study of the natural world in the Classical period of the seventeenth and early eighteenth centuries classified natural phenomena according to visible similarities. He then records how Classical classification gave way to the concept of 'species'. Thereafter, towards the beginning of the nineteenth century, the focus on species gave way to a new conceptualisation – that of 'life' itself – and, with this novel concept, a new science emerged, that of 'biology'. Biology was distinguished by its preoccupation with the internal organisational attributes that were, then, said to distinguish a thing as a 'living' thing. By the end of the twentieth century, along with many other sciences of living systems, biology began to understand what it is to be a living thing in terms of what Stuart Kauffman, theoretical biologist at the Santa Fé Institute, and holder of the 'MacArthur Genius' award, called 'self-organization' (1993, 1995, 2000). This understanding of life, as a self-organising and 'emergent being', is what reinforces the intimate correlation of risk and biopolitics.

Biopolitics, nonetheless, also changes according to changes in the technologies through which 'life processes' are made transparent to knowledge and, thereby, manifest in the world of appearance – including those designed, like marketing, for example, to regulate the external behaviour of biological entities, as well as those, like molecular science, designed to intervene into their internal structures. As Foucault taught (2007), the genealogy of the biopolitics of security is deeply implicated in the emergence of these and other modern developments, such as the rise of statistics, the specification of laws of probability and so on. Biopolitics of security thus constitute a plural and changing nexus of power/knowledge. As Foucault also insists, in many places too numerous to cite, knowledge and power are different enterprises. Thus, they cannot be conflated, but they are nonetheless also intimately allied. How that alliance works cannot be assumed. It has to be tracked historically, and established analytically.

Recalling briefly how, and why, biopolitics of security differ from more traditional rationalistic and geopolitical accounts of security, inasmuch as they revolve around life and its properties rather than sovereign territoriality, the present chapter explains why risk is also a natural corollary of the biopolitics of security. In pursuing that explanation, the chapter also explores the development and deployment of the risk technologies of contemporary finance capital. The 'risk' made, and traded, there, through such financial instruments as futures, derivatives and hedge funds, for example, as well as in the secondary insurance markets, is now not only a pervasive governmental technology. It is also one of the principal currencies through which contemporary biopolitics further extends the universalisation of utility as it enacts the transactional economic logic that Foucault first identified in his 1970s lectures (2003, 2007, 2008). Contemporary risk begins to display some of the key properties of money, very much in the sense of a new form of currency, to enable further transactional and combinatorial exchanges between entities made fungible and manifest in terms of the measure of their exposure to contingency.

Biopolitical life: transaction, contingency, risk

In the process of accounting for the biopoliticisation of security, and enframing risk as one of its technologies, this chapter offers a somewhat novel historicisation and theorisation of both risk and security. Indeed, the two necessarily go together biopolitically, since all problematisations of security revolve around a referent object, and all security technologies revolve around changing understandings of the properties of that referent object. Life, as a biological phenomenon, is the referent object of biopolitical security technologies. Biopolitical security technologies, therefore, reflect and seek to regulate according to the prevailing properties of the transactional existence of what, in his early lectures on biopolitics, Foucault called *être biologique* (2007). Among these are principally now said to be the transactional features of utility, fungibility and mutability, as well as 'circulation', 'connectivity' and 'complexity'. It is through these novel neo-baroque devices that life now largely makes its appearance in governable form. Circulation, because that is how biological life propagates and reproduces itself: 'Life, at its core, depends upon autocatalysis, that is reproduction' (Kauffman 2000: 15). Connectivity, or radical relationality, because biological life is understood to be the creatively propagating sum of its combinatorial transactions – 'Molecules are combinatorial objects', 'An autonomous agent is a relational concept' (Kauffman 2000: 44, 53) – and, the more things circulate, of course, the more they become transactionally connected. Complexity, because biological being is not a 'complicated' entity whose change can be plotted and projected in a simple linear fashion (Jacob 1989; Kauffman 1993, 1995, 2000). From the beginning of the nineteenth century onwards, albeit in significantly changing ways, the life of living things has been regularly said to be complex, subject to non-linear transformation and change (Jacob 1989; Kauffman 1993; Dillon 2005, 2007; Dillon and Lobo-Guerrero 2008). These, then, are the new truths, and so also, therefore, the new rules of life, that the Life of species existence brings to presence as governable materiality.

What characterises, and combines, all these generic properties of species existence is, however, radical contingency. The point applies to Machiavelli in one way as it does to the molecular revolution in another: 'It is central to Schrödinger's image, and more so to contemporary molecular biology, that the sequence of (DNA nucleotide) bases can be arbitrary' (Kauffman 2000: 26). In *Security, Territory, Population*, Foucault (2007) also explored how early biopolitics revolved around the aleatory properties of population. Just as populations are aleatory, however, so contingency is said now to be generically characteristic of organic life as biological existence. No divine providence, or historical teleology, explains the account of human existence as biological existence. It has no necessary cause external to it, and it follows no predetermined course. It underwrites itself, and appears as the continually emergent sum total of its transactional encounters with the contingent conjunctures that are said to characterise the life of living things. The space of that transaction is thus said by widely influential authors (like Kauffman, for example, who calls it the biosphere) to be 'incalculable, non algorithmic, and outside our capacity to predict (Kauffman 2000: x). 'Emergence and persistent creativity in the universe', says Kauffman (2000: x) with ontological directness, 'is real'. 'No one', he continues, 'designed and built the biosphere. The biosphere got itself constructed by the emergence and persistent co-evolution of autonomous agents' (Kauffman 2000: 3). Note also how the transactional economic space of biological being is additionally conceived, here, as a space of freedom, and that that space of freedom does not simply consist in rational calculative exchanges. It is said to consist in infinite complex combinatorial transactions that propagate diversity and organisational complexity, a newly baroque posing of the problematisation of its government as an infinity of finite things.

The centrality that contingency has thus acquired in the life of human existence understood biologically does not deny the observations of biological laws, or the systematicity observed in biological organisation and function. Kauffman is a biological scientist. Along with many others, he therefore understands these laws differently: as the laws of self-organisation and emergence that arise in a transactional space of emergence, itself undetermined by any transcendental necessity. Nor is the contingent, as it emerges here, understood as mere chance. In this latest twist to the ancient debate over contingency and necessity (Vuillemin 1996), as taken up in the early baroque and beyond into the neo-baroque of our own times, the modern account of the contingent does not simply elevate contingency over necessity. Contingency, itself, becomes a necessity in the novel domain of calculability through which the taming of chance is integrally involved in a new game with time. It is the infinity of factically finite time, itself freed from transcendental goals and laws, that helps institutes the rule of truth and truth of rule of contingency towards the end of the twentieth century, because of the ways in which it opens a space of probematisation whose conditions of possibility result in these kinds of risk-based conditions of operability.

Thus, it may appear that, in taming chance, you may tame time. Moreover, tame time and you may tame the future. Tame the future and you may, finally, secure a being – human being – whose very existence is temporal. However much it may characterise the modern, commanding time by securing the future is a long-standing human desire. At the beginning of the twenty-first century, it has, however, been given a quite distinctive gloss by life scientists like Kauffman, as well as by other sciences, such as those of the managerial and complexity sciences that now share cognate understandings of the emergent character of living things. Its impact upon information and communication technologies, as well as on military strategic discourse, is also now well documented (Dillon and Reid 2001; Dillon 2004). Kauffman (2000: x), for example, maintains that we cannot 'pre-state the configuration space of a biosphere'. Complex adaptation within it produces unpredictable, in principle infinite, self-engendering and self-organising diversification. It is this idea, specifically, that inspires the Unisys epigraph on security that heads this chapter, and the Unisys website brilliantly illustrates the spread of such neo-baroque thinking into the management sciences and the new sciences of security. Securing the future no longer simply entails the taming of chance, through the avalanche of printed numbers and technologies enabled by the rise of statistics (Hacking 1975; Bougen 2003). In addition to securing through simulation, prediction, pre-mediation, prevention and pre-emption (De Goede 2008; Aradau and van Munster 2007; Grusin 2004), life as emergence is increasingly now subject to an additional and quite distinctive moral and behavioural economy of existence. In a novel baroque market, it seeks to regulate possibility beyond probability (Amoore 2013).

Here, in the biopoliticisation of the securing of the life of emergent entities, virtuality is increasingly more important than actuality, the law of emergence similarly more important even than the laws of simulation and projection. Biopolitically speaking, at the beginning of the twenty-first century, biological being as emergent being is enjoined to secure itself through securing its future by experimental participation in the engendering and unleashing of its own, baroquely conceived, emergent potential. While allied to other ways of taming chance, risk technologies are also now deeply implicated in this novel biopoliticised securing of the life of emergent entities. And so: 'What if?', the Unisys website goes on to say in heightened neo-baroque tones, 'What if security wasn't a cage? What if, instead of keeping things out it let amazing things in? What if, it made you bolder, more ambitious, and enabled you to accomplish more than you ever thought possible? What if security unleashed your full potential?' 'Relish Change', it proclaims with typical commercial and baroque hyperbole. 'Get more control by controlling less' (Unisys 2007). If the infinity of finite things is baroque, this preoccupation with time and the future in all its infinite becomings and ramifications is baroque in the classical as well as novel ways in which Octavio Paz, for example, has defined the it analytically (1990). 'Neither history nor myth,' says Paz, 'the time that is coming [the time that is here] will be defined by the here and now ... The present doesn't project us into any space beyond any haphazard here, other-worldly eternities, of abstract *paradisios* at the end of history. *It projects into a medulla, the invisible centre of time; the here and now*' (1990: 138–9, emphasis added). This very cyclone of transformation and change, to which the biopolitics of twenty-first century is baroquely committed, is a direct result of what Paz also defined as 'the acceleration of history'. I will return to this acceleration – an 'acceleration despite itself' as Carl Schmitt observed in relation to the imperial drive of the United States (please see Conclusion to this book).

Contingency thereby becomes the epistemic object for biopolitics of security in the twenty-first century, inasmuch as it characterises the understanding of human life as an emergent and creative entity to whose promotion and development biopolitics are now committed. Extrapolating beyond Foucault's conclusions in *The Order of Things* (1989), it is the contingent, together with its so-called laws of emergence, rather than Man, that draws the life sciences together today. In that respect, contingency operates as a broad field of formation for the biopolitical security practices of emergent beings, which beings are increasingly also recognised to be both post- as well as extra-human. Thus, contemporary biopolitics also displays post- and extra-human concerns with the equally emergent life said to be displayed, for example, by artificial and cybernetic, as well as animal and viral, beings (Doyle 1997, 2003; Hayles 1999; Bingham and Hinchliffe 2008; Hinchliffe and Bingham 2008).

This point also teaches us something absolutely fundamental about how such biopolitics of security secure – indeed, about the kind of security that is pursued biopolitically today. Contingency cannot be excluded from life, since the very understanding of life that operates here, biopolitically, is that contingency is itself constitutive of what it means to be a living thing. If life, understood as biological being, is to be secured, such life cannot therefore be secured from contingency. Wrapping it up, to preserve life from the vicissitudes of its contingent being, will assuredly kill it off. Biopolitically, life has to be secured through contingency (Dillon 2007). If biopolitical security technologies are to succeed in their promotion of life as contingent biological being, if biopolitics are to ensure that life as emergence lives through acting out its infinite potential, then biopolitical security technologies, like risk, will not – because they simply cannot – secure life prophylactically. If life is, therefore, to be secured biopolitically, its securitising will in large part be conducted through the regulation of life's exposure to, and its productive and profitable exploitation of, contingent happenings and effects, including not only those occurring in nature, but also those that follow from the independent actions and interventions of biological being itself.

Biopoliticised security cannot, therefore, function by providing security merely in terms of safety or freedom from harm. Biopolitically, security is a game in which human life, as emergent biological life, must be promoted and secured through the regulation and fructification of its defining transactional properties and capabilities. This observation is not surprising. Indeed, it helps explain many current security preoccupations with networks, development, and local and global flows of every description, as well as the whole tenor of contemporary concern with the novel dangers that arise out of our close-coupled global civilisation, and the proclamation of radical uncertainty that now characterises security policies of almost every description.

It therefore follows that biopolitical security practices must somehow deliberately allow for the transformation and change – indeed, are distinguished by their desire to cultivate the very capacity for adaptive emergence – that living contingently is now said to require of all biological things. Biopolitically speaking, in other words, survival is increasingly said to be subject to the 'laws' of emergence. To survive is to change. By this is meant not just any change, not simple quantitative change, but qualitative change in the very nature of the living thing itself. This accounts for how transformation has entered into contemporary security discourses as a key term of the art. In the context of life, understood biopolitically, for an autonomous body that survives through adapting to the many contingently driven changes that duration through time as transactional and emergent space requires of it, transformation simply is a security strategy. This also explains why transformation has become as much of a concern to network-based or network-enhanced military strategic discourse, for example, as it has to corporate behaviour and contemporary management science (Farrell and Bird 2010). These and many other domains of contemporary governance – global liberal governance and development included (Duffield 2001) – share in this current bio-political zeitgeist (Dillon and Reid 2001; Dillon 2002). Or, as Kauffman (2000: 240) notes in an unintentionally revealing observation: 'If you are lucky enough to be in the survivable regime, you can survive by being adaptable.' What Kauffman gives away is the prior fact that it is the construal of 'the regime' as 'the survivable regime' that dictates that adaptation is required of autonomous agents, if they are to survive. Life, here, is then construed as that set of strategies capable of adaptively enduring through an economic space of combinatorial transaction. Once more, however, Kauffman unintentionally reveals something about this space. He also calls it a 'configuration space'. It shapes the autonomy and the agency of the very agents said to be operating autonomously within it. The biosphere, or econosphere as he later calls it, is not simply a biological economy. It is a biological economy operating very much also as a moral economy – except that its ethic masquerades as a life science rather than a political or moral science.

Biopolitically speaking, contingency is therefore not simple uncertainty. It names a particular kind of modern ontology – or understanding of the real, as the earlier Kauffman quote indicates – in which contingency is construed as the condition of emergent possibility, not simply for any kind of existence, or any kind of 'life', but for a life that is also 'free'. From this biopolitical perspective, then, living things are not simply defined as living things inasmuch as they are free to transact (molecular biology tends to conflate this transaction with 'information' exchange; Kauffman does not). Biopolitically, emancipation is the freedom of a transactional life dealing in the very combinatorial contingencies of its being in order to improve the conditions of its emergent existence, and prosper transactionally as a living entity. Kauffman puts it with typical, and revealing, biopolitical force: 'Agency was spawned.' Such transactional freedom might be said to derive from the natural attributes and emergent condition of biological beings – that is what it is to be a living thing in the transactional space of the biosphere, which knows no law other than that of emergence. This, emergent biological entities themselves creatively enact. For Kauffman 'an autonomous agent must be an autocatalytic system able to reproduce and able to perform one or more thermodynamic work cycles' (2000: 49). Note also the correlation, here, of Kauffman's understanding of what is more accurately termed a biopolitical, rather than a biological, autonomy, with work cycle. Here, he unintentionally reinforces Foucault's observation that economy provides the basic grid of intelligibility for the biopoliticisation of modern politics.

Where human beings are concerned, however, such combinatorial transactional freedom is, of course, operationalised through power/knowledge relations: legislated and regulated through the socialisation, education, training, discipline and norms that increasingly take the form of the multiplication of liberal rights to transact, and thereby become subject to the extension of utility. This increasingly now finds its expression in the laws of self-organisation, mutability, substitutability and emergence. In this account of biopolitical life, it is important to observe that freedom is no longer simply a matter of the will; the biopolitical agent is no simple rational calculative subject whose passions have been transferred into interests (Hirschman 1997). This is a biopoliticised agent whose reason and interests have been translated into autocatalysis. As a biological requirement of emergent existence, freedom becomes a biopolitical obligation, if one is in fact to be enfranchised in this neo-baroque polity as a living thing at all. Such a biopolitical franchise now admits non-human and non-organic entities into the category of the living, just as it regularly demotes, or excludes some human categories, from within the realm of life (Doyle 1997, 2003; Hayles 1999; Dillon and Lobo-Guerrero 2008).

Among the self-governing biopolitical technologies of transactional freedom are those technologies concerned to take the measure of contingency and trade on future transactional outcomes through specific kinds of statistical calculation and probability analysis. In other words, risk calculations. Pat O'Malley recorded how such transactional 'prudence' was first forcefully allied with political morality in the liberal utilitarianism of the nineteenth century. 'In this way,' O'Malley observes, 'uncertainty [first] came to represent to liberals not simply the "incalculable", but a specific set of techniques centred upon foresight: particularly contract, prudence and enterprise … For such liberals … uncertainty made them free' (O'Malley 2007: 7). Little surprise, then, that, in conflating biology and economy, Kauffman inevitably arrives at a biological expression of utility as well: 'Let happiness or the economist's utility, become "rate of reproduction", hence fitness. Let increased happiness become "increased rate of reproduction", hence increased fitness' (Kauffman 2000: 78).

In thus enframing risk biopolitically in relation to the contemporary understanding of life as emergent being, the emphasis thus far has been on the 'bio' of biopolitical. It is precisely that preoccupation that leads Kauffman (2000) to discover, in newly crafted form, liberal utility, economy and political economy 150 years after their first invention, discovering them alive and well and governing the biosphere. What I want to emphasise, instead, is the 'political' of biopolitical, for biopolitical enframing of risk is a political and not a sociological, anthropological, managerial or (except in the widest transactional sense) economic, much less biological, enframing of risk. This difference makes a difference.

From the wider political analytic adopted, here, the political is concerned with the generative principles of formation, institutions and practices – *pace* Foucault, both macro and micro – that instantiate a regime of power/knowledge relations, an allied social order, a system of distributive justice and a political morphology (an account of what it is, in politically salient terms, to be or become a human being). It is a commonplace of political thought that the 'social' of sociology, the 'anthropos' of anthropology and the 'subject' of economics do not precede politically instantiated orders of power relations (Pocock 2003; Lefort 1988). Neither, in fact, does Kauffman's emergent adaptive life. These co-evolve, one might say, with politics in and through regimes of governmental power/knowledge, encompassing the commercialisation and military strategic framing, for example, of contemporary technoscience.

How does this affect the analytic of risk offered here? In the following way: if economics, management and strategic science understand risk to be a universal phenomenon, something that is discovered by improved forms of knowledge and calculation (Bernstein 1998), sociology and anthropology understand risk to be a function of the demands or properties of some social or anthropological formation (Douglas 1992; Douglas and Wildavsky 1982; Beck 1992). Here, it is neither. Rather it is a governmental phenomenon. Precisely because it has become so biopolitically dominated, governance these days increasingly takes place through strategisations of the contingent, including greater refinement of risk awareness, risk-based analysis and risk-trading. Risk thus operationalises the biopolitics of emergent life through the commodification of contingency, now widely assumed to be constitutive of what it is to be a living, emergent, transactional being. As a device that directly operationalises our contemporary biopolitics of security, risk is therefore central to the neo-baroque governmental politics of formation that now dominate our world. And, there, it is traded to infinity.

If political modernity thus arises for Foucault, as it does for many others, as a generic crisis in the locus of political rationality and the efficacy of technologies of governance brought on by the dissolution of the soteriological project of the Christian Church, and its associated political institutions of Empire and Kingship (Kantorowicz 1997; Koselleck 2000 [1988]), modernisation, too, is a project and not an accomplishment. While modern political rationalities draw their licence, their very *loci* of power/knowledge, from a wide variety of reference points other than that provided by the Christian God, the search for autonomous regions of being from which political rationalities may be derived did not stop with 'population', 'civil society' or 'economy'. The realities of the infinite government of finite things and processes takes many finite forms. As an autonomous region of being, the referent object of life, too, changes. The early modern history and subsequent development of risk technologies lie embedded in the political, social, cultural and economic turbulence of these historical changes as well, and in the emergence, especially, of the new political rationalities and technologies of biopolitical governance revolving around the contingency of life understood emergently.

Risk: the contingent is no accident

Pascal may be said to be the first to have invented probability, in order to help him and his friends win at gaming (Bernstein 1998; Hacking 1975; Campe 2012). But, those who first compiled statistics, or experimented with new financial devices such as annuities did so to raise monies not only for gamblers and republics but also for kings. The United Provinces under de Witt financed their war against Spain through selling annuities. Without the demographics to guide their marketing, de Witt's annuities failed. But as they did so, they emphasised the need for new mathematical means of translating contingency into risk, and of turning numbers into newly effective foreign and domestic policy, as well as profit (Van Der Heijden 2006; Hendriks 1851; Poitras 2000; Clark 1999).

Similarly, when English forces under the Duke of Marlborough began the destruction of the French bid for European and transatlantic hegemony in the eighteenth century, they were successful not least because the English war effort was significantly better financed through the public credit made available through English commercial invention and success. Whereas such credit helped underwrite the success of the British bid for a global commercial empire, French failures in these respects and others fatally undermined France's *ancien régime* (Pocock 2003: 571; Brewer 1988; De Goede 2005; Brantlinger 1996; Hont 2005). Policy and profit are, of course, currently now also allied through credit in the form of risk in today's global financial capital markets. Consider the recent subprime lending crisis and financial catastrophe it helped precipitate.

The history of the emergence and operationalisation of risk has, therefore, always been situated at the intersection of capital and rule. Rule seeks to secure governability. Capital seeks to profit. Risk combines the two in posing and securing subjects of self-rule not simply in conditions of uncertainty, but in terms also of measuring their exposure to contingency financially. The extension of utility via pursuit of profit may assist the securing of governance. The securing of governance may assist the pursuit of profit. If orders of capital accumulation and orders of governance have a history, and if those histories are closely entwined, the same applies also to contingency and risk. Contingency has a history as well (Vuillemin 1996; Heller-Roazen 2003; Campe 2012). Specifically, today's 'risk' is a historical governmental phenomenon of intensively capitalised regimes of biopolitical security (O'Malley 2006).

As Robert Deuchars (2004: 30) has also recently observed, modern conceptions of risk calculation can in part be traced to an ontological shift that began with attempts made by scholars and activists in Renaissance Europe to question the moral and secular power of the established Church. Most notably, Machiavelli opened up a new field of thought and practice in which hitherto uncommon questions related to human control over nature, free will, providence, reason and progress were used to confront the orthodox thinking of the Church.

The space of problematisation that these developments opened up for new political rationalities and governmental technologies was, therefore, also related, among other things, to this changing problematisation of uncertainty, or chance, as well as to the life of the species, that has progressively characterised regimes of modernisation (Deuchars 2004; Dillon 2007). Precisely because population is characterised by the aleatory, however, it is this biopolitical focus on the referent empirical object of species being, in the form of population, that began the process of moving the contingent centre stage as a field of formation for modern biopoliticising governmental technologies. In so doing, population lent itself to governmental regulation, among other things, through technologies of risk. All this, Foucault documents, was intimately connected as well with the wider circuits of circulation and production of every kind that were associated with urbanisation, capitalism, the market and the evolution of economic theory (2003, 2007, 2008). Towards the end of the twentieth century, however, especially with the advent of the understanding of living things in terms of emergence, contingency seems to have graduated from being one biopolitical condition of biopolitical government among others to ontological pre-eminence, among contemporary biopolitical security technologies.

Risk has, however, also been conflated with the occasion of danger or the threat of loss (Beck 1992). This account of risk helps fuel the hyperbolicisation of security and fear that seems to characterise our contemporary politics. What is regularly omitted in such accounts of risk is, nonetheless, the equally important point that risk is simultaneously also associated with the occasion for gain. Our entire global civilisation revolves around the nexus of profit and loss that informs risk, and that sophisticated and inventive forms of risk analysis and risk packaging, in their turn, now govern. Much critical social and security analysis nonetheless continues to emphasise the threat aspect – real, manufactured or purely imagined – of risk.

Strictly speaking, however, risk is neither the occasion of danger nor the occasion for profit. Risk is simply the commodification of exposure to contingency calculated through the generalised measure of probability. Risk commodifies contingency by first making it calculable and fungible. Events and eventualities are allocated probabilities, a generalised measure of account, then correlated with their projected outcomes and given a score. People take a chance on that score. In simple terms, they bet.

Calling this a bet does not diminish what is involved. Quite the contrary, it more honestly expresses what precisely is involved. Our military and political leaders are perfectly aware of the fact that their Schmittean miracle of sovereign decision-making is more a bet than a miracle. They do not so much 'decide' in the Schmittean sense. They gamble. The architect of the contemporary military strategic doctrine of network-centric warfare, Rear Admiral Arthur Cebrowski, was among the very few to admit as much:

> Placing a Few Big Bets – These are big jumps. These are the things that will change a military service, change the Department of Defense and maybe even change the world. Some might argue that this is not what the DoD does, but they are wrong because the organization has already done this in the past.

> Global Positioning System satellites are a prime example. Its advent changed the military, changed the department and changed civil society. Another is the American military's ability, led by the U.S. Army, to 'own the night'.
>
> (Cebrowski n.d.)

What once helped substantially govern our forms of life in the developed world was the way in which risk was transacted through its socialisation and regulation according, for example, to law, to cooperative institutions, social norms or religious proscription. Progressively, however, and throughout all walks of life, such devices have increasingly become subaltern to that other single most powerful device for transacting risk, namely the free market.

The market has become massively more important in the second half of the twentieth century, of course, than at any other time in history, and across more of the globe than heretofore. It has become so in ways that have also transformed the very nature of capital, money and risk itself – to say nothing of Life and life. The changes are very generally recognised to be not simply quantitative, but also qualitative.

We should, therefore, follow Foucault in asking about the governmental effect of changes in the mode of capital accumulation and circulation 'between men and things' that occurred during the second half of the twentieth century. And we should do so in the way that his *The Birth of Biopolitics* lectures did when addressing the economic/governmental changes that followed the Second World War, by adding the changing order of 'securitisation' furnished by the dynamic and transformative ordering of capital itself, in which 'technologically amplified circulations of capital, information and forms of technology are now pushing aside the older economies of industrial production' (LiPuma and Lee 2005: 404). In other words, we have to do what Foucault taught us to do and ask: what is the governmental effect of the transformation of capital through which we have been living for the past thirty years, and how are we newly secured through those governmental changes? What we find when we ask these questions is that risk is no longer one device for offsetting certain specific exposures to contingency like accident, injury or death through the long-standing historical device of insurance. Risk provides a generative field of formation for the many orders of governance through which we are now increasingly secured. Risk is, in short, the predominant rule governing the conduct of conduct in the now deeply and pervasively capitalised biopolitical security practices of the developed world.

Underwriting security

Risk therefore comprises, today, a political rationality, and an assemblage of governmental technologies. The political rationality of risk proclaims an ontology of radical contingency. The technologies of risk seek to provide means of navigating that contingency, avoiding loss and seeking gain. The conduct of conduct in developed societies, from individual behaviour related to savings, mortgages, pensions and healthcare through to government policy in all areas, is increasingly governed by such risk analysis.

Those who proclaim the reality of risk presuppose a radical ontology of contingency, but what risk traders do is actually make risk. It is they who do the labour of calculating and commodifying contingency for the purpose of enacting a market in risk trading. Through such means, they calculate the probabilities of future events, buying and selling options and policies in relation to those projected events. Signing up for these, risk-takers run the risk of making a profit or incurring a loss. Such underwriting security does not differentiate inside from outside in a discursively organised play of friend/enemy or self/other. Neither does it immediately inscribe a social or political identity – populations and risk pools do not constitute a people in the usual political and cultural uses of that expression. Its domain of production is not that of identity but contingency, although it may then also lend identity to subjects of risk. It renews the circulation of exposure to contingency, and its allied topography of uncertainty, through the master trope of differential calculation. It does so in what two analysts call 'emerging circulatory forms [that] are fracturing the history of capital' (LiPuma and Lee 2004: 423). Here, a general theory of relativity in relation to exposure to contingency appears to be emerging in the form of the abstraction and universalisation of risk associated with the financial changes of the last thirty years or so. Underwriting security is precisely this assumption of risk (LiPuma and Lee 2004; Bryan and Rafferty 2006; Poitras 2000).

Underwriting does not, therefore, inscribe or ventriloquise a political subject of security through a complex discursive process of cultural and political inscription. Calculatively commodifying contingency, underwriting security assumes a risk and makes a pledge – the contractual act of underwriting as such – a pledge to expose oneself to and accept certain options or certain future conditions and outcomes. In this, it does of course both subject and subjectify, but the material cultural processes and outcomes are different from either those of geopolitics or of the writing of security.

Parties in exposure to contingency are thereby contracted. But, this is not the social contract of high modern political theory through which subjects and citizens are said to be secured in the enjoyment of their natural rights or traditional freedoms. This freedom is neither social nor political. The freedom is transactional, specifically the freedom to trade in one's exposure to contingency, regulating oneself according to how well or how badly one is able to capitalise that exposure. The contract is one of mutual but not equal liability. Underwriting, therefore, enacts a relation. But, the relationality is comprised of a specific liability, a liability entered into for advantage – although, of course, it may not turn out to afford one. It thereby also enacts a distributive political economy of who gets what, where, when and how. That economy includes a distribution of exposure to contingency, its fears and dangers, as well as a distribution of material benefits.

Whatever else might be said of the ontology of a contingent universe, then, 'risk' does not exist 'out there', independent, as it were, of the computational and discursive practices that constitute specific risks as the risks that they are. Risk is a carefully crafted artefact, as much as it is also an unintended consequence, not least of such risk making. Risks are thus baroquely created, circulated, proliferated and capitalised in a whole variety of ways. Underwriting secures self-governance through contingency by employing technologies of risk. This, then, is not exactly the risk society explored by Ulrich Beck (1992), whose account of risk is almost wholly preoccupied with cataloguing the growth of man-made dangers. This is risk as specific kind of truth-telling apparatus – what Foucault would call a mode of veridiction – of a governmental technology that enacts a world of self-regulating subjects bound in continuous calculation and commodification of their variable exposure to contingency. An infinity of finite risk operates, here, as risk becomes a dominant biopolitical technology.

Risk rules

The capitalisation associated with risk is classically that which now obtains in the world financial markets. It translates exposure to contingency into a measure, value or asset that can be transacted through free markets, both to offset exposure to contingency and to capitalise on it to make a profit. Insurance is one traditional device by means of which such securitisation via capitalisation takes place. Insurance, too, is being transformed by the changes to which I have just referred (Lobo-Guerrero 2011, 2012). The 'derivative' is a new financial device for doing more or less the same thing. Its origins are perhaps not as ancient as those of insurance, but in a sense its principle remains the same as that of insurance. It, too, deals in exposure to contingency calculated and commodified through risk. Its introduction has also transformed our world.

Insurance and derivatives, alike, therefore seek ways to offset exposure to contingency through the operation of a generalised system of real-time accounting of exposure to contingency. In the process, they have transformed the very character of global exchange and circulation, approximating it ever closer not only to comprehensive but also to real-time transaction of risk. They are doing this not simply in terms of the transformation of calculation and information. They are doing it in terms of what they measure.

What matters, therefore, is not simply that everything is being calculated and that everything is thereby becoming information, as many traditional accounts of global technologising from Heidegger (1977) to the present day maintain. What matters is what is being measured, and what the information gained is information about.

It seems banal to remind ourselves that, just as there are different ways of classifying and measuring things, and the history of calculation itself is also involved in these changes, there are also different things that are subject to classification and measurement (Elden 2006; Bowker and Starr 2000). But, the banality needs to be recalled. In a governmental order instituted by risk, the regulative ideal is that everything be measured in relation to everything else in terms of its exposure to contingency. What securitising according to risk does, therefore, is measure any aspect of anything in terms of such exposure. It is a market in the infinity of finite contingencies reliant upon risk analysis to furnish the positivity required to do the trading. In that way, any thing or any combination of things, real and virtual, may become measurable in terms of risk. As LiPuma and Lee (2004) argue in their persuasive account of the globalisation and abstraction of risk in today's financial markets:

> Capitalist social relations are no longer mediated only by labour but by risk, because these new financial instruments assume that particular forms of risk, no matter how existentially incommensurable … can be aggregated as an abstract form, susceptible to and determinable by mathematical calculation, combined within a single derivative, and then distributed to speculators.
>
> That is how the order of things has become the order of risk, and how risk has become the rule for the conduct of conduct in a new governmental order of securitising.
>
> (LiPuma and Lee 2004: 126)

While the implications of these developments – including their impact on what we understand the very nature of money as such to be – remain an object of academic dispute in academic journals, the global impact of derivatives and hedging on the order of capital accumulation as a whole is not in doubt. More than any other technological device, it is derivatives that are abstracting and generalising risk (LiPuma and Lee 2004; Bryan and Rafferty 2006).

The expansion of the derivatives market since the 1970s has transformed global economics and the nature of capital itself. What is especially interesting and important from the perspective of this chapter, however, is that derivatives have in the process also transformed the very range and register of governing through risk as well. Risk is no longer merely one way of dealing with exposure to contingency, by calculating and commodifying specific contingencies, as, for example, in the insurance business: third-party cover, accident, illness, injury, death, robbery and so on. Risk has become a general unit of social as well as of financial and managerial account (Power 1997, 2004, 2005, 2007).

The abstraction of risk as universal unit of account

> The financial community first envisions a situation as risky and then abstracts the element of risk from the intertwined social, economic and political circumstances that defined that situation. The identifiable risks are then moved to a conceptual space in which they may be considered independently of these circumstances.
>
> (LiPuma and Lee 2005: 414)

If risk technologies in their modern form were first enabled by Pascal's original account of probability in the seventeenth century, and by what Ian Hacking called the avalanche of printed numbers that characterised the eighteenth and nineteenth centuries (1982; see also Daston 1988), what LiPuma and Lee (2004) describe as their current 'universalisation' was enabled by the Blacks-Scholes formula that was invented in 1973. Its rapid utilisation was, in turn, enabled by the information revolution that followed digitalisation (Leslie and Wyatt 1992; Pryke and Allen 2000).

The universalisation of risk, nonetheless, also owed its advent to the ending of the post-war economic settlement, the abandoning of the Bretton Woods agreement, and the globalisation of finance, trade and manufacture that followed from those historical economic developments – key aspects of the very governmental transformations also addressed by Foucault in his *Birth of Biopolitics*. Taken together, the revolution in commercial and governmental power/knowledge of the last thirty years has transformed risk from one management device, and form of calculation, among others into what has many of the features of a universal system of account and a new order of governance. Risk, here, governs by continuously actualising the virtual as capital, thereby seeking to translate circulation into pure liquidity (Mackenzie 2007). But, it cannot be emphasised enough that such economic transformations are intimately allied with governmental transformations, and that the two together are always involved in crisis of governmentality and economy.

To paraphrase Simmel, substituting 'risk' for 'money' in ways that accord with the conflation of risk and money, which also appears to have been taking place, we might say that 'there is no more striking symbol of the completely dynamic character of the world than that of risk; and that, when risk stands still it is no longer risk'. Or, that risk is nothing but 'the vehicle for a movement in which everything else that is not in motion is completely extinguished' (Simmel 1991: 510–11). Such movement is the movement of time itself, the medium through which we are exposed to contingency, the medium that risk takes as its very domain of calculability, the infinity of indefinite things and processes in their infinite emergence to infinity.

In the realm of modern accounts of contingency as risk, time is therefore no longer the realm of fate, Fortuna, luck or accident. Time's uncertainty is no longer wrapped in a circulatory or cosmological mystery. Nor is the mystery of circulation simply governed by the secularised grace of Adam Smith's 'hidden hand'. Time in circulation is baroquely rendered as the contingent. Risk translates the uncertainty of this time into an embedded and embodied domain of social experience and an everyday order of governance. To adapt a Joycean expression coined by Simon Critchley and Tom McCarthy, that precisely expresses its baroque character, it is risk 'which makes the round go a world' (Critchley and McCarthy 2004). Modern risk transactions now seem to operate as a currency of exchange by means of which the value of things is measured, and any one thing can be traded for any other thing irrespective of what LiPuma and Lee call their 'manifold existential differences' (2004).

As sociologists of money, from Simmel onwards, have long observed, money is not just an infrastructural device enabling rational economic decision-makers to produce, buy and sell. It is an extraordinary performative social phenomenon, governmentally enacting and institutionalising all manner of values and social relations, not least, for example, those of trust and fidelity, as it both separates and joins people, and indeed generations, across space and time. In that sense, too, operating as a unit of exchange and a store of value, as well as instantiating specific forms of social relations and individual subjectivity, risk also appears to perform some of the classic functions that money is said to perform.

Recall, then, that risk is a calculation of exposure to contingency. Such information is stored in the way that money also stores information, in the form of price: the price of the policy, schedule, derivative or future (Muniesa 2007). As an apparently new iteration of money, risk now fuels significant transformations of the global economy and of the very nature of capital and capital accumulation. A measurement of contingency, a store of information concerning exposure to contingency, as well as a store of value, a medium of exchange and a means of payment, modern risk is not simply, however, a social phenomenon or a social institution. One person's risk is another person's income. Operationally, it is a governmental phenomenon, a complex governmental technology. Its self-regulatory register, range and power has vastly increased in extent and in intensity during the course of the last half century, impacting on everyday forms of life, including people's education, housing, jobs and pensions, in ways that affect their everyday behaviour and very understanding of the real. Arguably, the contingent now rivals the social and the economic as yet another of those autonomised domains of liberal Life.

Paradoxical as it may sound, therefore, risk satisfies the desire for security by upping the ante, and thereby massively increasing exposure to contingency, which it further translates into new risks. It does this by engendering an exponential increase in the ways in which everything is open to being addressed, valued and measured in terms of everything else, an infinite extensions of finite substitutions, using exposure to contingency as the general measure or universal unit of account and exchange.

Conclusion

There is, then, no one securitising game in the West. Securitising comes in the form of many games. These games are heterogeneous, and the security game-board is multidimensional. This allows moves to be made in and across several heterogeneous dimensions at once. Such worlds of securitising baroquely coincide, conflict and meld. Hence the use of the term palimpsest to describe how they overwrite one another, compounding the problem of reading and interpreting their complexly intertwined discursive textualities. Their relation is a strategic and not a dialectical one. Their complex and baroque intertextual inter-articulation very much helps define and circumscribe our current political condition. This world of risk is therefore not a mere supplement to the traditional world of security; it is a different world. But, recall also that it is perfectly possible for different worlds not only to coexist but also to transact as well as conflict with one another.

The widespread popular and professional mistaking of risk nonetheless distorts our political and governmental understanding of risk as a form of rule. It mistakes it from the beginning and blinds enquiry into the nature and operation of risk. The mistake is a fundamental and costly one, for risk has become one of the single most important security technologies of the age, and to mistake its character – to take its rules of truth and truths of rule at their own face value, so to speak – is to mistake the way in which we have come to be constitutively governed and self-governed through risk and its governmental technologies.

The political stakes involved in the mistaking of risk therefore could not be higher. To mistake risk is to mistake the shift in the ontopolitics as well as the apparatuses of power/knowledge and technoscientific devices, by means of which Western societies are now governmentally secured: from the macro-calculations of geopolitical analysis – no matter how scandalously they misconduct their risk assessments (Iraq) – to the biopolitical micro-management of individuals, populations and its account of Life as such.

Rule through risk thus secures individuals and populations locally and globally by locating them in a general economy of the contingent in which 'the event' rather than 'the will' reigns supreme. What that means is that the very understanding of the real that underpins risk-based approaches to security, as well as the technologies that are used and the knowledge upon which these technologies rely, differs significantly from that presupposed by traditional security discourses.

'Will', for example, is what the traditional preformed calculating subjects of security are supposed to possess. 'Events' are that to which contingently thrown human beings are exposed, not least the very unaccountable event of factically finitudinal time itself. Calculating subjects look for causes. Risk compiles statistics that calculate probability, abstracting in the process, also, complex patterns of behaviour. Preformed subjects of traditional security discourses are preoccupied in Hobbesian fashion with preventing bad things happening to them, especially bad things dreamed up by other, supposed and similarly calculating, subjects. Risk analysis pools individuals into risk pools, seeking profit by speculating on future events to which such pooled individuals may be exposed. Such security practices do not prevent things happening to people or corporations; they provide opportunity for gain, or they compensate people for any loss they may incur, allowing them to continue to actively circulate in the general combinatorial and transactional economy of contingency formed by risk.

The preformed subjects of traditional security discourses are said to be individual or collective, thus posing a so-called level-of-analysis problem for security analysts. The ontology of traditional security discourses is anthropocentric, in that its political anthropology of the innate cupidity of possessive individualism presupposes the existence of preformed rational calculative subjects of will, in the way that risk discourse initially does not. Its initial condition is an evental contingency amenable to calculation rather than a causal nature populated by calculating subjects. Reflecting the early scientific biases of the modern age, the epistemological posture of traditional security discourses is disposed towards proclaiming universal causal laws. Reflecting the probability revolution that swept through Western science throughout the course of the twentieth century, risk discourses are conversely also now probabilistic as well (Krüger *et al.* 1987). The subject of risk is a member of a risk-pooled cohort.

Risk ontology is, therefore, contingently evental rather than anthropocentric. Its world is governed less by sovereign wills and more by contingent occurrences, albeit contingent occurrences are governed even more explicitly now by the polysemous play of the sign in the form especially, for example, of infinite pricing and price fixing (LIBOR). The subject of risk-based securities is also preoccupied with survival. But the very reality of 'survival' is understood differently, here, as well. Surviving becomes a different game. It comprises a baroque world in which dependence upon the capacity to pass out of phase with oneself and become something that one was once not, produces ever more ornate (complex) social and economic as well as military strategic orders of governance in the accelerating time of real-time decision: a successfully traded risk-based commodity, for example, in which subjectivity is emergent and adaptive, articulating itself through a differential calculus of exposure to contingency as the mode of its own self-governance. Here, Life is as much an ontological as it is a biological or an economic category. In each instance it is conceived, temporally, as the ability to forget whatever has preceded the present situation. This is a perspectival world of relative valuation, more rigorous and more remorseless than progressive philosophers once proclaimed the emancipatory promise of the ontology of the event to be.

Unisys's world is, therefore, a world that directly recalls Paul de Man's account of the baroque character of modernity: 'This ability to forget and to live without historical awareness exists not only at the animal level. Since "life" has an ontological as well as a biological meaning, the condition of animality consists as a constitutive part of man.' This is another insight, one might add, into the ways in which the strategic figures of Man and Life are sutured together in the conjunction of contemporary bio and geopolitics of security. The moment of heightened humanity that Unisys claims may be secured through embracing security becomes a moment 'in which all anteriority vanishes, annihilated by the power of an absolute forgetting'. Which annihilating power is the power that the will to secure desires without remainder. De Man continues: 'Not only are there moments when it [animality] governs his actions, but these are moments when he re-establishes contact with his spontaneity and allows his truly human nature to assert itself' (De Man 1971: 146–7).

Whereas time for the traditional security subject is history, typically with a motor (*telos* or final end), time for the risk-based biopolitical subject of security is evental. It is futurally dependent, but it has no end. And the event knows no laws other than that of contingent emergence. There is no motor to the event, only conjunctures. Increasingly complex and ornate patterns can be detected through distribution and probability analysis, for example. But, the patterns are mobile and mutable, because they are also directly affected by reflexive behaviour itself. Power/knowledge, here, is more experiment in design than architectonic. Risk does not so much seek to ground things and build great institutionalised architectures of power/knowledge on those foundations. In recognition of its more complex and non-linear understanding of the nature of the real, risk espouses the complex productive play of contingency more than building, adaptation more than essence, transformation more than stasis.

Political philosophy is classically concerned with who should rule, how they should rule, and what binds subjects into an order of rule. It is preoccupied, that is to say, with questions of taxonomy and of political authority, political legitimacy and political participation among people said to share a common bond, the bond of submitting to a sovereign, agreeing to live by a set of shared political rules, or a constitution. Risk bypasses the traditional concerns of political philosophy. It nonetheless binds, and does so with an imperial impulse:

> the bond between economic subjects is, if you like, non-local. The analysis of the market proves that the multiplication of profits will ultimately be brought about through the spontaneous synthesis of egoisms over the whole surface of the globe. There is no localization, no territoriality, no particular grouping in the total space of the market.
>
> (Foucault 2008: 293–4)

The brute fact is that we are thrown into historical formations of government and rule – their rules of truth and truths of rule – long before we become capable of making any sense of them politically. Government and rule is a form of historical 'accident'. It never proceeds, first, from first principles. It also always defies the grasp of first principles, because it is messy, heterogeneous, contingent and complexly dynamic – features radically inimical to first principles. Rule through risk therefore also demonstrates how far removed we are from the ambitions of political philosophers concerned with the essence of politics. To mistake risk is therefore also to mistake the task of political philosophy that now confronts our age. Indeed, to mistake risk is to overlook the extent to which liberal political philosophy in particular has been comprehensively overtaken by its own liberal governmental technologies. The philosophical stakes in relation to the popular and academic mistaking of risk could, therefore, also, not be higher today, for risk newly challenges the labour of political thought required to contest the rule of risk, in particular the biopolitics and the geopolitics of security of security that it now baroquely enacts.

Foucault invites us to question the changing orders of governance through which we are ruled on an everyday basis. But he insists that we also ask how rule falters. If the 'quasi-transcendentals' of contemporary biopolitical governance – the 'circulation', 'connectivity' and 'complexity' which characterise 'life' – are now understood to be radically contingent and emergent, and if, as a consequence, risk has become a universal feature of biopolitical governance, then it is possible to conclude with a final observation in response to this second injunction. It is contagion, rather more than sedition, revolution or resistance, which threatens the rule of risk.

Notes

1 Taken from the Unisys webpage at www.securityunleashed.com/ (accessed 16 December 2007).

Bibliography

Amoore, Louise (2013) *The Politics of Possibility: Risk and Security beyond Probability* (Durham, NC and London: Duke University Press).

Aradau, Claudia and Van Munster, Rens (2007) 'Governing Terrorism through Risk: Taking Precautions, (un)Knowing the Future', *European Journal of International Relations*, 13(1): 89–115.

Baker, Tom and Simon, Jonathan (eds) (2002) *Embracing Risk: The Changing Culture of Insurance and Responsibility* (Chicago, IL: University of Chicago Press).

Beck, Ulrich (1992) *Risk Society: Towards a New Modernity* (London: Sage).

Bernstein, Peter L. (1998) *Against the Gods: The Remarkable Story of Risk* (New York: John Wiley).

Bingham, Nick and Hinchliffe, Steve (2008) 'Mapping the Multiplicities of Biosecurity', in Andrew Lakoff and Stephen J. Collier (eds), *Biosecurity Interventions: Global Health and Security in Question* (New York: Columbia University Press), pp. 173–93.

Bougen, Philip D. (2003) 'Catastrophe Risk', *Economy and Society*, 32: 253–74.

Bowker, Geoffrey C. and Star, Susan Leigh (2000) *Sorting Things Out: Classification and its Consequences* (Cambridge, MA: MIT Press).

Brantlinger, Patrick (1996) *Fictions of State: Culture and Credit in Britain, 1694–1994* (Ithaca, NY: Cornell University Press).

Brewer, John (1988) *The Sinews of Power: War, Money and the English State, 1688–1783* (Cambridge, MA: Harvard University Press).

Bryan, Dick and Rafferty, Michael (2006) *Capitalism with Derivatives: A Political Economy of Financial Derivatives, Capital and Class* (Basingstoke: Palgrave Macmillan).

Campbell, David (1998) *Writing Security: United States Foreign Policy and the Politics of Identity* (Manchester: Manchester University Press).

Campe, Rüdiger (2012) *The Game of Probability: Literature and Calculation from Pascal to Kleist* (Stanford, CA: Stanford University Press).

Cebrowski, David (n.d.) 'What is Transformation?', Office of Force Transformation web- page.

Clark, Geoffrey Wilson (1999) *Betting on Lives: The Culture of Life Insurance in England, 1695–1775* (Manchester: Manchester University Press).

Critchley, Simon and McCarthy, Tom (2004) 'Money and Morality in *The Merchant of Venice*', *Diacritics*, 34(1): 3–7.

Daston, Lorraine (1988) 'The Theory and Practice of Risk', in Daston, *Classical Probability in the Enlightenment* (Princeton, NJ: Princeton University Press), pp. 112–87.

De Goede, Marieke (2005) *Virtue, Fortune and Faith: A Genealogy of Finance* (Minneapolis: University of Minnesota Press).

De Goede, Marieke (2008) 'Beyond Risk: Premediation and the Post-9/11 Security Imagination', *Security Dialogue*, 39(2–3): 155–76.

De Man, Paul (1971) *Blindness and Insight* (Minneapolis: University of Minnesota Press).

Deuchars, Robert (2004) *The International Political Economy of Risk: Rationalism, Calculation, and Power* (Aldershot: Ashgate).

Dillon, Michael (2002) 'Network Society, Network-Centric Warfare and the State of Emergency', *Theory, Culture & Society*, 19(4): 71–9.

Dillon, Michael (2004) 'Correlating Sovereign and Biopower', in Jenny Edkins, Véronique Pin-Fat and Michael Shapiro (eds), *Sovereign Lives: Power in Global Politics* (London and New York: Routledge), pp. 41–60.

Dillon, Michael (2005) 'Cared to Death: The Biopoliticised Time of Your Life', *Foucault Studies*, 1(2): 37–46.

Dillon, Michael (2007) 'Governing through Contingency: The Security of Biopolitical Governance', *Political Geography*, 26(1): 41–7.

Dillon, Michael and Lobo-Guerrero, Luis (2008) 'Biopolitics of Security in the 21st Century: An Introduction', *Review of International Studies*, 34(2): 265–92.

Dillon, Michael and Reid, Julian (2001) 'Global Liberal Governance: Biopolitics, Security and War', *Millennium – Journal of International Studies*, 30(1): 41–66.

Douglas, Mary (1992) *Risk and Blame: Essays in Cultural Theory* (London and New York: Routledge).

Douglas, Mary and Wildavsky, Aaron B. (1982) *Risk and Culture: An Essay on the Selection of Technical and Environmental Dangers* (Berkeley: University of California Press).

Doyle, Richard (1997) *On Beyond Living: Rhetorical Transformations of the Life Sciences* (Stanford, CA: Stanford University Press).

Doyle, Richard (2003) *Wetwares: Experiments in Postvital Living* (Minneapolis: University of Minnesota Press).

Duffield, Mark R. (2001) *Global Governance and the New Wars: The Merging of Development and Security* (New York: Zed Books).

Elden, Stuart (2006) *Speaking against Number: Heidegger, Language and the Politics of Calculation* (Edinburgh: Edinburgh University Press).

Ericson, Richard V. and Doyle, Aaron (eds) (2003) *Risk and Morality* (Toronto: University of Toronto Press).

Ericson, Richard V., Doyle, Aaron and Barry, Dean (2003) *Insurance as Governance* (Toronto: University of Toronto Press).

Farrell, Theo and Bird, Tim (2010) 'Innovating within Cost and Cultural Constraints: The British Approach to Military Transformation', in Terry Terriff, Frans Osinga and Theo Farrell (eds), *A Transformation Gap? American Innovations and European Military Change* (Stanford, CA: Stanford University Press), pp. 35–58.

Foucault, Michel (1989) *The Order of Things: An Archaeology of the Human Sciences* (London and New York: Routledge).

Foucault, Michel (2003) *Society Must Be Defended: Lectures at the Collège de France, 1975–76* (London: Picador).

Foucault, Michel (2004) *Abnormal: Lectures at the Collège de France, 1974–1975* (New York: Picador).

Foucault, Michel (2007) *Security, Territory, Population: Lectures at the Collège de France, 1977–78*, ed. Michel Senellart (Basingstoke: Palgrave Macmillan).

Foucault, Michel (2008) *The Birth of Biopolitics: Lectures at the Collège de France, 1978–1979*, ed. Michel Senellart, François Ewald and Alessandro Fontana (Basingstoke: Palgrave Macmillan).

Grusin, Richard A. (2004) 'Premediation', *Criticism*, 46: 17–39.

Hacking, Ian (1975) *The Emergence of Probability: A Philosophical Study of Early Ideas about Probability, Induction and Statistical Inference* (Cambridge: Cambridge University Press).

Hacking, Ian (1982) 'Biopower and the Avalanche of Printed Numbers', *Humanities in Society*, 5: 279–95.

Hayles, Katherine (1999) *How We Became Posthuman: Virtual Bodies in Cybernetics, Literature, and Informatics* (Chicago, IL: University of Chicago Press).

Heidegger, Martin (1977) *The Question Concerning Technology, and Other Essays* (New York: Harper & Row).

Heller-Roazen, Daniel (2003) *Fortune's Faces: The Roman de la Rose and the Poetics of Contingency* (Baltimore: Johns Hopkins University Press).

Hendriks, Frederick (1851) *Contributions to the History of Insurance and of the Theory of Life Contingencies, with a Restoration of the Grand Pensionary de Wit's Treatise on Life Annuities* (London: printed by C. and E. Layton).

Hinchliffe, Steve and Bingham, Nick (2008) 'Mapping the Multiplicities of Biosecurity', in Ali S. Harris, and Keil Roger (eds), *Networked Disease: Emerging Infections in the Global City* (Malden, MA and Oxford: Wiley-Blackwell), pp. 173–94.

Hirschman, Albert O. (1997) *The Passions and the Interests: Political Arguments for Capitalism before its Triumph* (Princeton, NJ: Princeton University Press).

Hont, Istvan (2005) *Jealousy of Trade: International Competition and the Nation-State in Historical Perspective* (Cambridge, MA: Harvard University Press).

Jacob, François (1989) *The Logic of Life: A History of Heredity; the Possible and the Actual* (Harmondsworth: Penguin).

Kantorowicz, Ernst Hartwig (1997) *The King's Two Bodies: A Study in Mediaeval Political Theology* (Princeton, NJ: Princeton University Press).

Kauffman, Stuart A. (1993) *The Origins of Order: Self-Organization and Selection in Evolution* (Oxford: Oxford University Press).

Kauffman, Stuart A. (1995) *At Home in the Universe: The Search for Laws of Self-Organization and Complexity* (Oxford: Oxford University Press).

Kauffman, Stuart A. (2000) *Investigations* (Oxford: Oxford University Press).

Koselleck, Reinhart (2000 [1988]) *Critique and Crisis: Enlightenment and the Pathogenesis of Modern Society* (Cambridge, MA: MIT Press.).

Krüger, Lorenz, Daston, Lorraine, Heidelberger, Michael, Gigerenzer, Gerd and Morgan, Mary S. (eds) (1987) *The Probabilistic Revolution* (Cambridge, MA: MIT Press).

Lefort, Claude (1988) *Democracy and Political Theory* (Minneapolis: University of Minnesota Press).

Leslie, J. and Wyatt, G. (1992) 'Futures and Options', in David P. Cobham (ed.), *Markets and Dealers: The Economics of the London Financial Markets* (Harlow: Longman), pp. 85–110.

LiPuma, Edward and Lee, Benjamin (2004) *Financial Derivatives and the Globalization of Risk* (Durham, NC: Duke University Press).

LiPuma, Edward and Lee, Benjamin (2005) 'Financial Derivatives and the Rise of Circulation', *Economy and Society*, 34(3): 404–27.

Lobo-Guerrero, Luis (2011) *Insuring Security: Biopolitics, Security and Risk* (Abingdon: Routledge).

Lobo-Guerrero, Luis (2012) *Insuring War: Sovereignty, Security and Risk* (Abingdon: Routledge).

MacKenzie, Donald (2007) 'The Material Production of Virtuality: Innovation, Cultural Geography and Facticity in Derivatives Markets', *Economy and Society*, 36(3): 355–76.

Muniesa, Fabian (2007) 'Market Technologies and the Pragmatics of Prices', *Economy and Society*, 36(3): 377–395.

O'Malley, Pat (ed.) (2006) *Governing Risks* (Aldershot: Ashgate).

O'Malley, Pat (2007) 'Uncertainty Makes Us Free: Risk and Rationalities of Freedom' (Keele University).

Paz, Octavio (1990) *Conjunctions and Disjunctions* (New York: Arcade Publishing).

Pocock, J. G. A. (2003) *The Machiavellian Moment: Florentine Political Thought and the Atlantic Republican Tradition* (Princeton, NJ: Princeton University Press).

Poitras, Geoffrey (2000) *The Early History of Financial Economics, 1478–1776: From Commercial Arithmetic to Life Annuities and Joint Stocks* (Cheltenham: Edward Elgar).

Power, Michael (1997) *The Audit Society: Rituals of Verification* (Oxford: Oxford University Press).

Power, Michael (2004) *The Risk Management of Everything: Rethinking the Politics of Uncertainty* (London: Demos).

Power, Michael (2005) 'The Invention of Operational Risk', *Review of International Political Economy*, 12(4): 577–99.

Power, Michael (2007) *Organized Uncertainty: Designing a World of Risk Management* (Oxford: Oxford University Press).

Pryke, Michael and Allen, John (2000) 'Monetized Time-Space: Derivatives – Money's "New Imaginary"?' *Economy and Society*, 29(2): 264–84.

Simmel, Georg (1991) *The Philosophy of Money* (London and New York: Routledge).

Unisys (2007); Unisys – Security Unleashed – Unleash Your Potential'. <www.securityun-leashed.com>.

Van Der Heijden, Manon (2006) 'State Formation and Urban Finances in Sixteenth and Seventeenth-Century Holland', *Journal of Urban History*, 32(3): 429–50.

Vuillemin, Jules (1996) *Necessity or Contingency: The Master Argument* (Stanford, CA: Center for the Study of Language and Information).

5

POLITICS OF TRUTH AND PIOUS ECONOMIES

> Act in such a way that you give the gods something to do.
>
> (Kafka, quoted in Lambert 2008: 60)

Introduction

This chapter is devoted to an analysis of the changing character of Christian pastoral power after the Reformation, and of the contribution it made to the biopolitics of 'development' through European colonial exploitation from the late seventeenth century, and the recruitment of so-called faith-based organisations (FBOs) by global liberal government from the late twentieth century onwards. The particular focus for the first is the evangelical Lutheran Francke Foundation instituted in Brandenburg-Prussia in the late seventeenth century, and, for the second, the Churches so closely implicated in US and British development–security complexes from the late twentieth century. These were not, of course, the first attempt to mould Church and Churches into an *instrumentum regni*, but their exercise took place, henceforth, in relation to the novel exigencies of factical finitude (Maravall 1986).

The religio-political nexus of the modern age

From the seventeenth century onwards, all political and governmental debate in the North Atlantic basin addressed the question of the effectiveness as well as the legitimacy of rule – 'effectiveness' in relation to securing the peace and prosperity of domestic populations as a concomitant of securing the best form of government: effectiveness, that is to say, in relation both to delivery as well as legitimacy (Maravall 1986). Somewhat contrary to Foucault – and what Foucault calls the royal questions of power, concerning the derivation and exercise of sovereignty – the division of powers, and the organisation of the offices and revenues of the sovereign (cameralism), always also addressed how the favoured form of government would improve, renew or advance the material conditions of the subjects of the sovereign in the process also of figuring how best to awe the population. The issues of the commonweal – whether it was construed as subject, citizen or people – and of its improvement and renewal as a matter of the good order supplied by rule, were always addressed as well. It was axiomatic that the right form of government was one that would necessarily, also, lead to the improvement of the conditions of populations. How people were governed was necessarily correlated, then, with how well they were governed – in respect of provision not only for their basic material means of subsistence, but also for their religious and civic training. There were none that argued for the rule of liberty – equality and representation – alone, or indeed, for that of divine right or tradition. Spectacle was consequently, also, always related to performance of course as well. No performance, no spectacle.

Royal questions of rule were therefore also tied up with what Foucault called the conduct of conduct, more generally, and with welfare of the commonweal, however the commonweal was figured. Hence, from the very early days of modern political rule, sovereigns were always also in the market for strategic ideas about how to govern populations and, in particular, how to govern them in ways that also made them more productive, since productivity was closely associated with domestic peace and tranquillity at home, as well as geopolitical power abroad. Economy was integral to governance, just as spectacle was integral to rule. Baroque political artifice addressed both domains. The two went together. They still do: not least in terms, especially, of the development of financial systems in the late twentieth as much as in the eighteenth century, and of novel financial mechanisms strategically driven to finance war as much as they were commercially driven to promote trade (Brewer 1988; Brantlinger 1996; De Goede 2005). Indeed, once more, geopolitics, pastoral power and biopolitics could hardly be disengaged from one another. So far as the modern state was concerned, Churches could be strategic partners in these enterprises, for the Christian Churches were also in the business of rule, particularly in respect of governing the conduct of conduct. But they also had a long track record of being violent strategic adversaries and competitors with one another as well as with the state.

The point about the evangelically inspired Christian Churches, in particular, such as the Lutherans and contemporary Pentecostal evangelicals (to which reference is made later), and in addition to the Jesuits, is precisely that their activities were never confined to the spiritual realm. Their work has been part of the vast reformation of Christian confession in which a foundational issue as old as St Augustine recurrently comes to the fore. It now arises under the novel conditions set by political, economic and commercial as well as scientific modernisation and development associated with factical finitude, as well as religious reformation. Modernity did not therefore settle the problem of the religio-political nexus of truth and rule. It refigured it in ways that continue to exceed its enframing by secularism, as well as by fundamentalism, something that baroque politics captures very well.

Critics write as if there is, however, something integral to the Foucauldean analytics that inspire this chapter, which somehow 'marginalises' the interrogation of forces at work in the modern transformation of truth and rule that Foucault nonetheless also frankly acknowledged – such as those, for example, of dominance, brute extraction and primitive accumulation. Others similarly claim that Foucault's interrogation of discursive practices of biopower, so influential in critical development studies, and of the orientation of power around the properties of species being rather than the subject, is somehow not 'material' (Wilson 2012); as if such *dispositifs* were not themselves brutally material, and experienced in brutally material ways. It is a curious account of materiality that excludes such practices from the realm of materiality. Foucault certainly did not (cf., for example, the later lectures in *Society Must Be Defended* [2003], where he discusses the racism and genocide implicit in biopolitics). What crass materiality is it that cannot also recognise that the very processes of materialisation always already signify more and less than they materialise? Baroque politics as well as science played a very material part in this transformation of the materialisation as well as the understanding of the materiality (Maravall 1986; Gal and Chen-Morris 2013a, 2013b). Rules of truth and truths of rule – the rites of material political manifestation – make things happen, and confer a concreteness materially exceeding 'materiality', by manifesting a world (Schwartz 2008); albeit a world of exploitation, injustice and self-aggrandisement, as well as spectacle and glorification.

Much of this debate comes down to what Foucault called the politics of truth, and to how thinkers come to be read through them (2007); in other words the complex processes of pedagogy, politicisation and de-politicisation, as well as of contrived appearance, bound up with the correlations of truth and rule. Suffice to say that Foucault never attempted to address all aspects of the manifold of modern power relations, that he continuously stressed that he was not trying to develop a general model of power, and that he did not say that biopower succeeded sovereign power. But he did draw an important distinction between domination and power, observing that, while domination of course existed, one of the features distinguishing the plural and diverse, as well as changing, manifold of modern power relations was the way in which these worked through the specified freedoms of modern subjectification. More the pity, then, that he did not pursue the baroque spectacle of sovereign power with which he opened *The Order of Things*.

Novel ways of disciplining bodies were an additional aspect of modern power relations that he observed. Another aspect seemed focused less on sovereignty, individual bodies or the freedom of subjects than on the properties of species existence, for which *dispositifs* of power relations Foucault borrowed and adapted the term 'biopower'– which did not thereby mean that sovereignty, individual bodies or subjectification ceased to be points of application for power relations. Population was a specific – material – point of application for the operationalisation of biopower. Foucault continually, also, repeated the classically observed point that government had to take into account the nature of the thing being governed. In his insistence on the correlation between knowledge and power, he also noted how knowledge provides an account of the nature of things that makes them amenable to governance; albeit rendering things knowable in different ways opens them up to being governed in different ways as well. One of the ways in which the politics of truth takes place is the operationalisation of power/knowledge.

In no place, and in no way, then, did Foucault proscribe the interrogation of domination, of the *dispositifs* of sovereign power or indeed of any other manifestation of power relations – such as that, for example, of structural concentrations of state and corporate power capable of exercising forms and forces of domination with global and psychic reach unattainable in previous centuries (Butler 1997). If other aspects of modern power relations have been 'marginalised' – for whom, where, and how? – such marginalisation is not warranted by Foucault.

Foucault may be more fairly criticised, however, for the European and Franco-centric parochialism of his account of the development of power relations to which he devoted his attention (Stoler 1995). Many of the distinguishing features of modern power relations to which he drew attention emerged simultaneously, also, in European colonies and plantations, and in England, Holland and the Americas as much as they did in France and Germany (Rusnock 2002; Cassedy 1969). This also included early and brutal forms of biopower in Ireland as well (Petty 1690; Buck 1977, 1982). So what? Equally, he might be criticised for neglecting the contribution to novel technologies of government, made by the revision of Christian pastoral power that followed the break-up of Christianity into rival Christian Churches, albeit his last lectures were precisely concerned with providing a genealogy of Christian truth-telling as well as truths of rule. Moreover, except for his reports on the Iranian Revolution (Afary and Anderson 2005), Foucault had little to say about the mutual implication of religion and politics so characteristic, even then, in the 1970s, of modern politics nationally and internationally.

There are many gaps and lacunae in Foucault's work that are not mandated by that work but which leave an opportunity, instead, to extend the kind of questioning that he pioneered, such as that specifically, here, of the politics of political truth that emerged in a factically finite world with which modern 'faith-based organisations' have also had to contend. Many such gaps can also be addressed simultaneously, since the revision of modern religious power relations had as much to do with the participation of the Christian Churches in the geopolitical and biopolitical spread of commercial imperialism, racism and colonial conquest, as it did with sovereign spectacle as well as their participation in shaping modern scientific and social mores.

Since Constantine (306–337 AD), at least, Christianity had formally defined itself in relation to the secular authorities and demands of imperial power, as much as it had in relation to the soteriological and spiritual power of its God. Similarly, from the seventeenth century onwards, the Reformed Christian Churches of Northern Europe, in particular, defined themselves as much in relation to the newly emerging national and international power relations instituted or patronised by the modern state – such as disciplinary and biopower – as they did in relation to the reformed interpretation of their God and his worship. Christian Churches, along with other religions, now construed by neo-liberal governmentality as 'faith-based communities', continue to do the same today.

God had only ever been one pole, or point, of problematisation and application, however, for the development of Christian rules of truth and truths of rule. From the very inception of Christ's ministry, let alone Paul's revolutionary interpretation of it, so influential among Pentecostal evangelical churches in the United States and sub-Saharan Africa today (Marshall 2010; Chan 2011; Robbins 2010), secular political authorities had always been another pole for Christian politics of truth. The very differentiation of the secular from the spiritual was, itself, originally, a theologico-political accomplishment of the Christian differentiation of eternity from finitude, and of God from Man (Augustine 1984; Markus 1988). The secular is, then, a religio-political trope, and modern secularisation a reference to how a swing from the religious to the secular became a defining feature of political modernisation, in what, nonetheless, continued to remain in many respects a deeply baroque religio-political nexus of truth and power relations.

As the rule of truth underwent profound technoscientific and religious transformation during the course of the sixteenth and seventeenth centuries, so also did corresponding truths of rule. Similarly, as new truths of rule emerged from the dynastic and increasingly national rivalries of early modern states, so also were new rules of religious as well as political truth enunciated – unleashed as much as by the rise of trans-oceanic commerce, enslavement and technoscientific revolutions in war-making (Parker 1988; Glete 2002), as by the break-up of the Holy Roman Empire's enframing of the political order of the *Res Publica Christiana*.

From the very advent of modernisation, Churches responded directly to the challenges of modernisation and development materially as well as theologically, in the realm of knowledge and social relations as much as in the realm of conscience, liturgy and observance. The reformed Churches were especially committed to the development and improvement of the material conditions of individuals and populations. For the Christian Churches, material welfare was to be reconciled with redemption. For newly emerging states and forms of governance, material welfare had to be reconciled with sovereignty and its geopolitical ambitions, as well as the governability of individuals and populations. There was a strategic intersection, here, between the Churches and the State, but a strategic intersection that by no means assumed an identity of purpose. Atlantic ideals of republican, and later also liberal, political virtues were not to be conflated with religious ideals of public morality (Pocock 2003; Skinner 1998, 2008; Joyce 2003). But they could be, and regularly were, strategically allied.

Such strategic intersections were and remain, however, contingent and not guaranteed. Thus, by the end of the twentieth century, the civic ambitions of Atlantic liberalism and republicanism had been widely reduced to a baroque governmental arabesque of security politics. More than that, government has undergone a biopoliticisation in which both local and global poor, reduced to threatened populations exposed to uninsurable risks, are widely regulated and policed now through the provision of military intervention and emergency relief. In the process, the high republican and liberal ideals of civic virtue, along with later social democratic ones, have given way extensively to regulation by public, often populist, morality, and media-disseminated religiosity and spectacle, backed by bloody displays of military potency. In this regulation through public morality rather than civic virtue, whose single most popular point of application is the local and the global poor, notably in Africa, the Churches have found the basis for a new strategic alliance with liberal and republican regimes high on hyperbole, low on delivery and in thrall to capital and military more than political virtues (Patterson 2011). For many influential military strategic thinkers of the last twenty-five years, capital and martial virtues have amounted also to much the same thing, which accounts, in part, for why they share the same popular bio-philosophical discourses of networks, complexity, the emergency of emergence, event and resilience (Dillon and Reid 2009; Osinga 2007).

One indicator of the evisceration of republican and liberal political culture is, indeed, this new strategic interdependence with religion via the very designation of confessions of faith as faith-based organisations. Thus renamed, religions are rendered more governable. That way, also, the discursive practices of government are more readily able, than they once were, to recruit religion to the rules of truth and truths of rule currently propagated by neo-liberal governance locally and globally. Hence, FBOs now widely administer US and UK funded aid and development programmes in biopolitical regulation of the local and the global poor, 'eventally' focused on complex health, food and security emergencies widely assumed to inspire terrorism, AIDS, famine, drought and civil war. Throughout these developments, while the wealthy are most heavily armed and bellicose, it is the poor that seem to inspire most fear.

The generative principles of formation operationalising these programmes thus translate civic virtue into emergency relief, the city into battle space and the pursuit of the good into event-driven kairologically-inflamed fear; fear pervaded by the anticipation of cascading catastrophic breakdowns of subsistence, law and public order nationally and internationally. Some of their specific points of application are food and health and, especially in the United States, the unwed ethnic woman and the emergency ward of the local public hospital. Politics has become conflated with government; citizenship with membership of endangering as much as endangered populations; and civic virtue with more intensely regulated sexual as well as governmental conduct. Once celebrated as the privileged locale of a vibrant democratic political culture, cities are widely construed as conflict zones.

As a complex mutating assemblage of truth and rule, the religio-political nexus of the modern age thus yields analytically to the Foucauldean enframing of the politics of truth. As the seventeenth century drew to a close, in few places was early modern transformation in the politics of truth undergoing more enterprising revision than in the new baroque university town of Halle in Brandenburg-Prussia. Religious ferment mixed with pedagogic as well as scientific and political change. Indeed, it was hard to differentiate between them – since rule was at issue in science, teaching and religion as much as commerce and profit. Salvation was always already at issue in the newly emerging discourses, institutions and practices of politics, government and rule. When are they not?

If the religious, and other, revolutions of the sixteenth and seventeenth centuries had, as Foucault felicitously observed, stimulated a crisis in the conduct of conduct *tout court*, in few places was that opportunity to re-problematise seized with such pedagogic, entrepreneurial, artistic, scientific and religious zeal than in Halle, and by the Lutheran evangelical August Hermann Francke (1663–1727). Francke's zeal was matched with equal, if contrary determination, however, by a former co-religionist ally, the jurist and political theorist Christian Thomasius (1655–1728) (Hunter 2006, 2007; Hunter *et al.* 2007). The vicious personal conflict that erupted between them was a microcosm of the stakes at issue religio-politically in Europe at that time, and throughout its burgeoning colonies overseas as well. Namely, how were faith and civics to accommodate one another to the dual ends of glorifying God and governing Man? If this was an old problem, the revolutions of the sixteenth and seventeenth centuries had blasted open the ways in which – via the arabesque political ornamentations and intersections of sovereign and biopower – it was reproblematised, addressed and readdressed, from then on. That microcosm is a good place to begin this further reflection.

The pious developmental economy of the seventeenth century

Civic decorum

The jurist and Protestant theologian Christian Thomasius contested the metaphysical connection between divine and human reason championed by the evangelical Lutheran Pietist August Hermann Francke, and other religious enthusiasts. In keeping with his drive, less to eliminate faith than to separate faith and reason, Thomasius denied this continuity. His argument was theologically twofold. First, that the damage done to man's faculties at the 'Fall' meant he could not hope to derive the norms of natural law by exercising a reason similar to that informing natural law. Second, he made great use of the theological category of *adiaphora* or 'indifferent things' (*Mitteldinge* in German). *Adiaphora* referred to all matters said to be neither commanded nor forbidden by God. They were thus irrelevant to the question of salvation, for which God's commandments could then, Thomasius argued, be reduced to three: to love God; to love one's neighbour; and to have contempt for oneself as a creature of passion prone to disorder and in need of correction, both civic and Christian. As a result, many of the things that the competing confessions had declared to be essential, and over which so much violence had occurred – such as church liturgies, sacraments, the Trinity, transubstantiation and so on – could, thereby, be declared to be matters of moral indifference. That way they could be construed as matters of 'Christian freedom' or political regulation (Hunter 2007; Hunter *et al.* 2007). This nonetheless still left quite a lot to fight about. Compare, for example, how Thomasius derived these new truths of rule from his reinterpretation of the Christian rule of truth, with the history of government via the Eucharist in the medieval history of Church and State.

According to Henri de Lubac (2006) and Michel de Certeau (1992), two changes had a profound impact on the way religious life was instituted in the medieval period: the specialisation of the elite class, including the priesthood, and the organisation of the Church into a more centralised body whose theology was more technical, doctrinal and professionalised. These developments were an extension of a complex set of developments impacting the Christian rule of truth and truth of rule from the advent of the first millennium onward, including, in particular, a radical shift, also, in its attitude towards war and its relationship to the military classes, as the Church itself progressively mutated into an imperial institution (Tellenbach 1948). Of first, and most enduring significance, were, however, the Gregorian reforms that, directly stimulating, required a crushing military as well as liturgical and doctrinal response to the many religious movements and heresies that sprang up across the whole of Christian Europe from the twelfth century onwards (Grundmann 2005; Tellenbach 1948). Thus, the Fourth Lateran Council (1215) made sacramental practice 'the instrument of a campaign to free Christians from the grip of the first large popular heresies, autonomous communal movements, and growing secular powers'. In the meantime, the Eucharist in particular was also given a strategic function: to consolidate the Church, by positing not just the equivalence but the identity between mystical reality and the visible, and by making this depend upon the hierarchical authority and dispensation of the Church (De Certeau 1992: 85–6; De Lubac 2006). Kantorowicz (1997) documented how the model of the Eucharist was later appropriated to sacralise Kingship, underwriting it with a theologico-political rule of truth/truth of rule. For Thomasius, as much as for the Scholastics that he, like Hobbes, abjured, theology was politics and politics was essentially a theologico-political problematic, however much the Christian God had by that time become a *deus absconditus*, cutting more slack in the universe for Man (Blumenberg 1985). But, it was one construed differently, and the construal required a re-engineering of the institutions of both Church and State as well as the training of the functionaries that staffed them.

Thomasius' campaign was, of course, part of a much broader struggle that was taking place throughout Europe and in European colonies overseas. On one side were those for whom the consolidation of a new form of territorial political sovereignty required the general exclusion of the Christian Churches from the exercise of civil power. On the other side were the clerical estates whose imperial rights, in particular, had permeated the exercise of such power, through the threat of blasphemy, heresy and witchcraft, for example, as well as of excommunication. The problem was compounded by the new interpretation of faith and worship introduced by the Protestant religion, particularly Luther's pronouncement of *sola fides*, salvation through faith alone. It compounded the problem because it also transformed the everyday conduct of conduct of adherents to Protestant articles of faith. Its new rules of truth demanded new truths of rule, and both were challenging to established secular authorities, although they could also be recruited to support newly assertive and ambitious princes and states.

In terms of his interpretation of the problematic of Christian truth, then, Thomasius' campaign amounted to a generic attack on the truths of rule and rules of truth of the confessional Christian state. It divided religion and government in ways that removed salvation from the domain of Church ritual, while removing political authority from the domain of salvation. The purpose was expressly that of developing new truths of civic rule that, rendering other clerical truths both theologically as well as politically illicit, served as a means, also, of revising clerical rules of truth in order to establish grounds for driving clerics from the juridical and political domain (Hunter *et al.* 2007: xv–xviii).

Thomasius' new 'civic decorum' (Walker 2002a, 2002b) directly opposed the new 'disciplinary' training and pedagogy for which Francke's schools were becoming famous. Through it, Thomasius not only conducted what was labelled a *decorumstreit* with Francke (Walker 2002a, 2002b), he deliberately formulated discourse and practices designed to open the way to training quite differently instructed legal and governmental cadres in the sciences of law, government and police (Hunter 2007: especially ch. 1; Dubber and Valverde 2006).[1]

This was no abstract doctrinal dispute. Construing it as such would entirely miss the point of what was at stake for the participants; stakes of which they were only too well aware. Neither was this only a dispute about constitutions, legitimacy and offices of state, although, of course, it was that too. According to his conception, in order to live peaceably in the post-confessional society whose very rules of truth and truths of rule Thomasius was seeking to reformulate – and whose 'decorum' or civic sociability he was trying to specify – it was necessary according to Thomasius for individuals to be trained in separating their personae as Christians and as princes, subjects and believers (Hunter *et al.* 2007). Thomasius' revision of the relationship between the theological and the political not only entailed a revision of the relationship between Church and State, from which neither were to emerge unchanged, it demanded a transformation of the lived embodiment of prince, subject and governmental functionary. The whole novel and embodied ensemble was interrogated and explored through the very baroque German dramas, or *Trauerspiel*, of the time (Benjamin 1998).

Evangelical discipline

Originating in the Protestant territories of south-west Germany in the 1670s, German Pietism was an evangelical Lutheran movement as much concerned with social as it was with religious reform. A 'Pietist' was someone who was affiliated with a reform movement first initiated by Philipp Jakob Spener in the 1670s. A long and deeply entrenched historiographical tradition has portrayed the proponents of this movement – especially Francke – as zealous, yet practical, Lutheran reformers who were forced to directly confront the ideals of early Enlightenment in conjunction with the state-building mandate of Brandenburg-Prussia.

In Foucauldean terms, Pietism was as deeply committed to the revision of the conduct of conduct as it was to revising the articles, liturgies and practices of Christian faith. The one project could never of course be divorced from the other, and so, with the reformation of religious faith and observance came a protracted revision of the pastoral powers of a Christian Church gone bloody and sectarian. The right of Pietist laypersons and clergy to preach as well as to teach and discipline, in the course of organising charity, religious observance and spiritual guidance, was contested by orthodox Lutherans, in the south German Duchy of

Württemberg, for example, and throughout several of the large German trading cities. Led by Francke, the Pietists nonetheless succeeded in establishing themselves in Brandenburg-Prussia, under the protection and patronage of Frederick II and Frederick William I. These allowed him to establish and develop a large charitable foundation and orphanage in the baroque town of Halle, in whose new university Francke also took up a university professorship.

The Francke Foundation rapidly became a form of global corporation with powerfully entwined religious and commercial trading links stretching from Muscovy in the east to the colonies of North America and the Caribbean in the west, and the Coromandel Coast of the Indian subcontinent in South Asia. Evangelical Christianity thus strategically aligned itself within the emerging political rationalities and governing technologies of state formation and empire building by powers of the North Atlantic basin, during the course of the late seventeenth and early eighteenth centuries, but it nonetheless independently, also, sought to pursue its own mission locally and globally; dispute over the issue of slavery in the colonies was an important illustration.

It would, therefore, be wrong to see Halle Pietists in particular as cultivating their identity in direct opposition to the civic enlightenment involved in the development of cameralist *staatswissenschaftlische* experiments in governance in the very state that gave the Pietists support and sanctuary. The relation was instead more complicated and reciprocal. The Halle method of educational reform, in particular, as well as the meaning that eighteenth-century Europeans were to attach later to philanthropy more generally, had a wide impact on the political and intellectual reformers of the period. The Pietist orphanage promoted Pietist pedagogy designed to introduce children to the conciliatory knowledge-making strategies of the first Berlin Academy of Sciences. These strategies championed the status of the 'heart' as an assimilatory juncture point for a reconciliatory form of governance and self-governance. It was the material expression of Halle Pietists' commitment to a 'third way', not so dissimilar in its ambitions from the general intent of the interfaith foundation established by former UK Prime Minister Tony Blair; albeit the Pietists were concerned to assimilate experience and cognition as well as theology, philosophy and voluntarism to absolutism, rather than global liberal governance.

Thus, according to one of the most detailed accounts of the work of the Francke Foundation:

> Pietist orphanages were both real and imagined spaces. They did and did not house orphans. Their founders were and were not Pietists. They were funded by everyone and no one (with the exception of God). They were and were not supported by the state. They were inclusive and exclusive. They existed to regulate both the environment and the soul. The educational programming in place within these sites aimed to create a world in which there was no difference between thinking and feeling, or between knowing and doing. Here the seemingly impassable fissure imagined as existing between the world of the mind and the world of the hand had been bridged and Cartesian dualism mediated. The founding of a Pietist orphanage, or application of its

> pedagogies in other institutional settings, signified the pursuit of a rigorous method of action and assimilation, a middle way. Following the example of the Jesuits, whose prowess as educators was widely known, Pietists cultivated the 'visual aptitudes' of children and held up the eye as the perfect conciliatory, didactic and edificatory medium. In the spaces touched by their methods, the eye and the heart became fused together into a single entity that observed, reconciled and loved.
>
> (Whitmer 2008: 1)

While religious, the Francke Foundation was nonetheless also committed to the improvement of populations pursuing the renewal of their governance as well as the salvation of their souls:

> Halle Pietists believed in the promise of helping individuals acquire 'visual aptitudes' through training in the mathematical sciences, which provided a methodological point of orientation for solving the problem of dissonance between competing regimens of truth – including competing confessional systems. Although, as Lorraine Daston has noted, 'the divergence, integration and transcendence of individual perspectives were the province of moral philosophy and aesthetics' in the eighteenth century more generally, in Halle, around 1700, this group of radical Lutheran theologians attempted to link moral philosophy and aesthetics to 'primitive Christianity' or the idea of a universal faith. Like Gottfried Wilhelm Leibniz, whose 'perspectival metaphysics' and preoccupation with harmony is well documented, Halle Pietists pursued a pansophic vision of universal harmony, justice or benevolence; only they institutionalized their version of this vision in an orphanage and placed it at the center (or heart) of a self enclosed 'city of schools.' They were participants, like Leibniz and so many others, in a culture of reconciliation, or what the noted Professor of Medicine at the University of Halle, Friedrich Hoffmann, called a 'culture of understanding,' and their conciliatory and vision oriented method was hugely influential precisely because of their ability to replicate it in new and existing institutional settings.
>
> (Whitmer 2008: 8)

Theirs was a knowledge-based as well as faith-fuelled experiment in social formation and reformation. They were:

> passionate inquirers, observers, consumers and assimilators of the cacophony of materials and methods so characteristic of knowledge making in eighteenth century Europe. Like Robert Boyle, Robert Hooke and other members of the Royal Society in England, they believed in the possibility of reconciling several forms of knowledge (and knowledge making) by exploiting the potential of middling objects and instruments. In doing this, they participated in the same sorts of conciliatory knowledge making as the polymaths they associated with: Ehrenfried Walters von Tschirnhaus, Gottfried

> Wilhelm Leibniz, Johann Christoph Sturm, Leonhard Christoph Sturm, Christian Wolff, Christian Thomasius, Friedrich Hoffman, Georg Ernst Stahl, Daniel Ernst Jablonski, even Frederick Slare (a former student of Boyle's and Royal Society member).
>
> (Whitmer 2008: 7)

Not unlike faith-based organisations today, missionising, medicine and teaching were its business. Thus, from the very beginning, it was trans-national and religio-commercial as much as it was religio-political. Similarly, also, it targeted public morality, the education and training of children and of future spiritual leaders. Its virtues were religious rather than political. But in as much as its pedagogy domesticated, as it religiously inspired, the Foundation contributed towards producing skilled, diligent and industrious subjects. This was no bad thing from the original governmental perspective of cameralism and police, or indeed later liberal and republican regimes, equally concerned with domestication and industry. These were to be pursued instead, of course, through the complex mixture of specified freedoms, discipline and biopolitics crafted by the development of liberal governance (Dean 2009; Barry *et al.* 1996). But such technologies could be strategically allied to the newly emerging pastoral powers of the evangelicals.

Inspired by the evangelical fervour of the European Reformation, Pietist beliefs were nonetheless often criticised as excessive and irrational. These created many surfaces of friction within the Lutheran Church as well as between the Lutheran and other Churches. In Tranquebar, on the Coromandel Coast of the Indian subcontinent, for example, evangelicals fought with other Protestant co-religionists such as the Anglicans as well as the feared Jesuits. In the states of the North American colonies, especially Georgia, leading Pietist pastors owned slaves and evangelical communities were generally divided on slave-holding, as well as related questions concerning the conversion of slaves and their attendance at Pietist churches. As the evangelicals came to eschew slavery, however, so ultimately they not only clashed with each other, but also with colonial governments and commercial interests, over the question of slavery (Stange 1968).

Despite Whitmer's observations, it is evident from her pioneering work, and other sources, that the relation of the Foundation to early modern politics was equally also characterised by mutual suspicion and contingent strategic alliances in respect not only of the Foundation's missionising work, but also the social reform pedagogy that it pioneered, the medicine that it marketed, its treatment of slaves, and its teaching on the spiritual as well as racial and economic status of slaves (Stange 1968). Inevitably so, the relationship of the Francke Foundation to the royal authorities in Brandenburg-Prussia, directly competitive also with that which Thomasius sought to cultivate, was one in which Church and State – though allied – were conscious of their differences.

More interesting still, a point that bears comparison with the operations of evangelical churches in the United States today, the Francke Foundation was commercially astute. Francke was an energetic self-publicist, an assiduous social as well as religious networker and tireless fund-raiser. Bankrolling its missionising by making and selling medicine, along with the Bible and other religious tracts, the Francke Foundation was as much a commercial, epistemological and pedagogical enterprise as it was a religious and spiritual one. Equally as much an experiment in ruling in the service of the Lord as absolutist *polizeiwissenschaft*, or liberal government and biopolitics developing alongside it, were experiments in the governance of states and populations. The Francke Foundation had, nonetheless, continually to ensure that, in serving the Lord, it was not simply also serving the state. In this respect, the situation for Christian Churches in the eighteenth and nineteenth centuries was little different, at least in principle, than that of so-called FBOs today. Their historical circumstances, of course, differ.

Whereas new cameralist *staatswissenschaftlische* and *polizeiwissenschaftlische* rules of truth and truths of rule proclaimed the sovereignty of the state, the Pietist Lutherans, along with other Christian Churches, ultimately proclaimed the sovereignty of God. The two had not been contingently reconciled since the Constantine conversion of the Roman Empire lost its Christian apologist in Eusebius, gaining a very different account of the relation between the City of Man and the City of God in the teachings of St Augustine. The issues ran deep and wide, neither were they confined to the problematic of sovereign rule alone. The conduct of everyday life, the very object of Pietist teaching and practice equally as much as it was for cameralist reformers throughout the following century in Prussia, or liberal governmentalists in France or the British Isles, was simultaneously at issue as well.

Governance, specifically self-governance, is required if sovereign will and the force of law is actually to work. Governance, specifically self-governance is equally necessary also to achieve salvation. That is why the Churches were in the business of rule equally as much as the secular authorities. For, every account of truth is accompanied by some corresponding governmental imperative that specifies how one should be governed or exercise self-governance in light of the truth proclaimed. Cameralists recognised this as much as preachers, or, indeed, liberals and republicans. Theorists of *polizeiwissenschaft* understood it just as well as physiocrats, political economists and liberal thinkers, like Adam Smith who wrote among other things, of course, on *The Theory of Moral Sentiments*. What differentiated them was the how of rule – the how of governance and self-governance – not *that* religious and political truth alike required there to be rule, and self-rule.

Operating within the strategic nexus of relations comprising new formations of power relations, Christian Churches, the Francke Foundation in particular, were therefore no mere dupes of the state, although they could operate as agents of the state. Neither were they dupes of commercial and economic exploitation, although they often also served the interests of colonisation and imperial conquest. Indeed their missionising was integral to the civilisational and racist rationales that provided the veridical and ideological underwriting to colonisation and imperial rule. The Churches, too, attended scrupulously to business in order to spread their message, and, in the process, it was often difficult to dissociate religion, as a business, from the business of religion. The point is that these dangers and difficulties were as much in evidence with Lutheran evangelicals, in the seventeenth and eighteenth centuries, as they are today with FBOs (Patterson 2011).

Whereas the Francke Foundation, correctly, feared being recruited into the cameralist drive of Brandenburg-Prussia and other European states, FBOs, Pentecostal evangelicals in the United States in particular, seem less circumspect today, however, being enthusiastic recruits to biopolitical and disciplinary neo-liberal targeting and regulation of the poor at home and abroad. (The point is the referent governmental object 'the poor' and less the constantly reproduced international distinction between inside and outside.) Then, as now, states and secular organisations also expressed suspicion of Church involvement in the improvement, renewal and developmental regulation of populations. But then, as now, the changing religio-political rule of truth and truth of rule of modernisation and development was open to strategic exploitation, by both sides.

The religio-political nexus of the modern age has, therefore, always been a shifting field of strategic formation and intervention, often as much adversarial as it has been cooperative. Whereas reforming political authorities were often deeply suspicious of religion, because it was held responsible for intractable doctrinal conflicts and protracted warfare, in a new expression of the long-established conflict between Church and State, the reformed Churches also remained institutionally suspicious of political authorities as well. For the vocation of the Church was to serve a higher authority. Having gained their freedom of belief and observance from Rome, the reformed Churches may have welcomed the protection of political authorities from Rome, and from princes loyal to Rome, but they remained conscious of the danger of conflating their mission with that of the new political authorities. If the Church had always been concerned with the material conditions of existence of its flock, it had always also been concerned with its own specific pastoral and disciplinary arts of governance as well. That said, it could not readily function if the secular authorities proscribed it, and the social order eschewed it. What was true then, remains true now.

In short, finding themselves in the midst of historical re-problematisations of the nature of nature, and thereby also of the corresponding re-problematisation of the very origin and nature of rule and law, temporal as well as spiritual, Church and State reformulated the how as well as the why of governance and rule (Daston and Stolleis 2008). And they did so in ways that made political authorities newly reliant – for the purposes, especially, of baroque display, spectacle and the arcana of statecraft – upon novel baroque political economies of the sign. Statecraft was signcraft, and no less liturgical and cultic, or preoccupied with glory, than it had ever been (Agamben 2011; Lambert 2008; Maravall 1986; Egginton 2010). It was no coincidence, either, as other students of the baroque have noted, that during a time of epochal upheaval early modern dramatic theory developed alongside of early modern political theory pioneering novel stagecraft as well as statecraft. Each was preoccupied with the formation of a modern subject, its psychology as well as its political anthropology, their *mise-en-scène*, plots, characters and tropes closely intersecting one with another (Harris, 2014). If this was the triumph of secularism, it was certainly not the triumph of a disenchanting politics. The rules of truth and truths of rule that underwrite twenty-first century government and politics, within as much as outside the North Atlantic world, are no less sacralised and cultic in their pursuit today than when Churches were more widely assured of their spiritual and temporal missions. Now, however, their societies of spectacle and their renditions of glory demand analytical attention extending beyond that once supplied by Guy Debord (1994) and towards that which analysts of the baroque and neo-baroque (Lambert 2008; Maravall 1986; Egginton 2010), in addition to, but in ways quite different from Agamben (2011), have indicated.

Indeed, the very exigencies and pressing urgency of how to rule were at least as important as the broader royal questions of the legitimation and formal constitutional structures of rule. Church and State, alike, experimented in the pursuit of everyday governance, as much as they did in formally constituting and legitimating, as well as allying, themselves as governors. That they therefore shared an interest in the development of new forms of governance via welfare, of which biopolitics was a variant, did not, however, mean that the problematisation of governance and rule arose for them in the same ways, was based on appeal to the same sources of rule, or was motivated by the same interests in ruling. These considerations applied to absolutists as much as they did to liberals and republicans.

The pious economy of the twenty-first century: faith-based organisations

By the end of the twentieth century, global rules of truth and truths of rule had changed dramatically from those obtaining at the end of the seventeenth century. While it is an exaggeration to say that sovereignty and *polizeiwissenschaft* were wholly superseded by neo-liberal *dispositifs* of power, characterised by the use of biopolitical technologies of governance, these do seem to have become much more widespread – a favoured mechanism of global governance. Meanwhile, also, discourses of modernisation and development became characterised by intense and novel interpretations of economy and security, of the economy of security, a response driven by a perceived generalised threat from the global poor to the globally advantaged in what had by then become a closely integrated global economy. Just as the Francke Foundation was implicated in the early modern emergence of new forms of governance and rule, so also are confessions of faith now directly targeted by neo-liberal governance in the guise of faith-based organisations recruited into their development–security complexes. This change was pursued most directly by the US and the UK, but it is a feature also of indigenous religio-political developments in sub-Saharan Africa.

In the US, the new strategic alliance between religion and politics was dominated by the growing strength of the religious right. This populist phenomenon was, nonetheless, also deeply implicated in the saturation of public discourse by a rabid free market ideology, relentlessly disseminated by corporate media, designed to rationalise and disguise a larcenous raid on wealth by corporate financial intuitions unprecedented in global, let alone US, history. Along with an astonishing transfer of wealth from the US 'middle' as well as its poorest classes to the already vastly wealthy (Piketty 2014), came a social agenda driven by Christian evangelical religious preferences and proscriptions. These twinned developments transferred also into the distribution of US aid and development overseas, the vast bulk of which now takes place through evangelical churches. By way of distinction, the recruitment of FBOs in the UK was driven by the perceived threat from its immigrant communities, increasingly typified as religious communities, and the widespread fear of radicalised elements of its Muslim youth, following the UK joining the US war on terror and invasions of Iraq and Afghanistan – 'strategic disasters' that British intelligence, among other governing institutions, noted, have materially increased the threat of terror attacks in the UK (Johnson 2014). Here, then, FBOs were recruited as part of a differently engineered development–security response to the securitisation and militarisation of British participation in the revision of the regulation of the global poor that followed the dissolution of the Cold War, and the global hegemony established thereafter by neo-liberal economic and geopolitical policies pursued also through key international institutions like the World Bank and the IMF as well as national security apparatuses (Dillon and Reid 2009).

FBOs in the UK

It thus came as little surprise to hear, in August 2001, in the midst of an historically unprecedented retrenchment of public expenditure, from reports in *The Guardian* leaking an internal UK Department for International Development (DFID) paper explaining that the new UK National Security Council had said that, 'the ODA [Overseas Development Administration] budget should make the maximum possible contribution to national security consistent with ODSA rules'. 'Although the NSC will not in most cases direct DFID spending', the DFID paper was reported as saying, 'we [DFID] need to be able to make the case for how our work contributes to national security' (Watt 2010). That additional national security value should be squeezed out of every aspect of the national budget when the UK fiscal military state found itself unable to sustain extensive borrowing from international capital markets to help fund overseas war, as it had been able to do so successfully since the financial innovations of the eighteenth century enabled it to defeat France in imperial competition for global hegemony, should come as no surprise.

But it did. Condemned by a radically discredited Labour opposition that had done more to securitise development than any previous British political administration, an anonymous New Labour source complained that: 'This document is deeply worrying, as it confirms the fears of many in the international development and humanitarian development community that the government plans to securitize the aid budget.' Labour had already largely securitised it. For, sometime during the course of the 1990s, as Mark Duffield has documented so well (Duffield 2001, 2007), a tight-knit, wide-ranging and more ideologically self-conscious alliance was formed between security and development among the liberal societies and states of the North Atlantic basin. It increasingly targeted, and more systematically incorporated, resourced and directed, Christian and humanitarian organisations pursuing health and educational as well as poverty and disaster relief programmes globally. The genealogy of development was the history of empire. The history of empire is as involved as that of security and development. Christian Churches played a notable part then as they do in the security–development complexes of the twenty-first century.

Hence, the UK government issued a White Paper on International Development in 1997, for example, committing the Blair government to prioritising support for development through enrolling businesses and trades unions, faith communities, black and minority ethnic (BME) communities and Diaspora groups. The aim was to further resource a development strategy already committed to linking poverty with war and the war–poverty nexus with threats to British national security and prosperity. The Building Support for Development Strategy (BSDS, April 1999), subsequently published in 1999, set out how this would be achieved. Whilst the overall policy context has developed over the intervening period, the BSDS has continued to set the direction of policy. The BSDS gave priority to education and the media, which were covered by seven of its objectives and the bulk of the funding. However, the strategy expressed a commitment to involve hitherto unreached parts of society through innovative organisational partnerships with businesses and trade unions, churches and faiths.

A short time later, Prime Minister Blair formally proclaimed the international strategy, latent within the emergence of the security–development complex, in a speech to the Chicago Economic Club in April 1999. Albeit prompted by the need to justify NATO intervention in Kosovo, Blair's Chicago speech detailed a wide-ranging doctrine of biopolitical liberal interventionism on the grounds that – given globalisation and the radical interdependence of world poverty and war overseas – global interdependence threatened national security and prosperity at home. Liberal war-making in pursuit of such intervention was justified on the grounds that it pursued 'values' not territory: 'This is a just war, based not on any territorial ambitions but on values.' Security, peace and prosperity were thus conflated, as were international boundaries as well as international distinctions between faiths, cultures and peoples. This recruitment drive was followed by other states as well. It was extended and intensified as part of the rollout of the war on terror after 9/11 in particular through DFID and USAID.

FBOs in the US

Towards the end of the 1990s, international and national donor agencies ranging from the World Bank and the United Nations to the UK Department for International Development were moving towards a consensus on privileging faith-based service providers of humanitarian aid and emergency relief (Cooper and Waldby 2014). In 1998, the World Bank convened a World Faiths and Development meeting which led to the creation of the World Faiths and Development Dialogue, an advocacy group now based at Georgetown University, promoting collaboration between faith-based service providers and international development agencies. Systematic Christian Right lobbying of the United Nations throughout the decade progressively committed UN agencies also to privileging faith-based provision of aid and relief (Cooper and Waldby 2014).

In May 2003, shortly before also launching the invasion of Iraq, George W. Bush announced a $15 billion President's Emergency Plan for AIDS Relief (PEPFAR) billed as the largest public health campaign in history to target a single disease. A five-year programme, it was designed to address the international AIDS epidemic in twelve sub-Saharan African countries and two Caribbean nations. With bipartisan Congressional support and the endorsement of the incoming President-Elect Obama as one of the success stories of the Bush administration the programme was re-authorised in 2008, providing a further $48 billion in funding over a further five years. Its scope was also extended to a further fourteen countries in the Caribbean (Wright 2009: 59–60).

From its very inception, PEPFAR was informed by theologically warranted prohibitions against extramarital, commercial and non-heterosexual relations that reversed previous public health strategies focusing on harm reduction rather than sexual regulation. Congress levied three conditions on the distribution of PEPFAR funds when it authorised the programme. The first required that at least one-third of all prevention funds should be spent on the promotion of sexual abstinence before marriage. Based on so-called ABC principles of Abstain, Be Faithful, use Condoms, condom distribution was treated as an absolute last resort. Second, FBOs were exempted from participating in prevention strategies they found theologically objectionable. Third, it applied a loyalty test. Any organisation refusing to state in public its opposition to prostitution and sex trafficking, thereby excluding agencies that worked directly with sex workers, would be ineligible for funding (Dietrich 2007; Ingram 2010). The effect of these and other provisions was to channel disproportionate amounts of funding to Pentecostal evangelical churches that were in any event already being privileged as the mechanism for the delivery of US aid (Clarke 2007; Gerhardt 2010; Hearn 2002; Patterson 2011).

The following year, Bush issued a further executive order creating an Office for Faith Based Community Initiatives within the US Agency for International Development (USAID). Its express purpose was to facilitate the channelling of aid projects to FBO service providers.

USAID subsequently published guidelines on 'Participation by Religious Orders in USAID programs' prohibiting discrimination against organisations that combined emergency relief with religious activities such as proselytising, ritual or scripture. Ostensibly designed to provide a level playing field for secular and religious organisations alike in fact it tilted the balance materially in favour of religious organisations since US or non-US NGOs providing information, for example, on abortion (and overwhelmingly secular) were of course already ineligible for USAID funding.

Whereas the UN-focused FBOs traversed the range of Christian denominations as well as Islam and Judaism, especially in the campaigns against AIDS, Pentecostal evangelicals dominate US FBOs. American Pentecostalism is a genre of contemporary evangelicalism deeply committed to an imminent messianic, eschatological and kairological interpretation of Christian faith. Salvation is close at hand but radically conditional upon observance of the moral law preparing its believers for the coming of the Kingdom. If the rule of truth of Christian messianism is that Christ will come again – and for the Pentecostals that coming is imminent – the truth of rule messianically is: be prepared for the coming. That spiritual preparedness does not simply thereby become a simple condition for the receipt of aid and emergency relief, strict observance in one's everyday conduct of conduct becomes the very way in which one qualifies through one's lived embodiment for that aid and emergency relief.

Confessions of faith thereby become protocols of rule. A world simultaneously of truth and rule is born here. Imminent anticipation of the Kingdom requires governance. Governance is delivered through the very conduct of conduct that anticipation of the Kingdom demands. What appears as Christian charity is delivered as Christian rule, rule delivered in sub-Saharan Africa, however, to communities that are nonetheless themselves widely governed by Pentecostal rules of truth and truths of rule.

Conclusion

> What I am doing is something that concerns philosophy, that is to say the politics of truth. So … what is involved in this analysis of the mechanisms of power is the politics of truth, and not sociology, history, or economics.
>
> (Foucault 2007: 3)

What is called the modern age thus arose out of a complex series of transformations in the orders of both truth and rule, as did its religio-political politics of modernisation, security and development as well as its revised promises of spiritual and political salvation. Political modernity was therefore not a period waiting to be discovered (Davis 2008). Modernisation and development, no less than security with which these are now also comprehensively linked, are not final states. Rather, their changing rules of truth and truths of rule are eschatologically and katechontically driven fields of baroque formation in continuous experimentation with government and politics: battle spaces of fragile and shifting strategic alliances comprised of complex surfaces of friction within and between new confessions of truth as much as new formations of rule.

From the outset, this contest extended across the entire field of human settlement and endeavour. Even as it was cultivated and pursued in a Europe hardly cognizant of its European-ness – its Christian-ness, however, deeply influenced by its encounter with Islam in general and the Ottoman Empire in particular – it was simultaneously and experimentally pursued eastward to Russia and China as it was to Africa, North and South America and the Caribbean. It was that differentiating encounter with other worlds which directly contributed to the formation of national and regional as well as religious and political accounts of 'Western', 'European' and 'national' identities (Todorov 1999; Anderson 1991; Pagden 1993). Similarly, also, did changing definitions of modernisation, security and development emerge.

In pursuit of their modern expressions of political and religious identity, the French became French, the English English, the Catholics Catholics and the Protestants Protestants. They did so, as much through murderous religious, political and commercial pursuit of one another by land and sea (then air) across the globe, as through the workings of putatively 'indigenous' forces of spiritual reformation, commercial industrialisation and wider socio-economic development. Competition for slaves, souls, territory and capital have to be distinguished from one another, but viewed as deeply complicit and intimately allied constituent elements of a breath-takingly violent, rapacious and transformative process – one operating in revised, but ever more powerful, forms, now threatening the very habitability of the planet, and with centres of power now lodged elsewhere than in the North Atlantic basin alone.

The racism that accompanied the very many different and mutating processes that claimed modernisation as their warrant was similarly central to them. It was not epiphenomenal. It was an integral and changing expression of the rules of truth that inspired modernisation as much as it was the truths of rule – of capture, exploitation and enslavement – through which these found their diverse forms of operationalisation (Eze 1997; Dillon 2008). It is perhaps only a slight exaggeration to say that many of the processes claiming to advance modernisation and development over the last three to four hundred years comprised doctrines of truth and formations of rule that had exploitation and enslavement – together with the displacement, dispossession and mass murder of entire populations – as their very *raison d'être*; leave aside the allied *raison* of their *raison d'état* (Davis 2002).

From the seventeenth to the twenty-first centuries, the 'missionising' of new religious, political, economic, racial and scientific rules of truth could not be divorced from the corresponding imperialising of their allied new truths of rule. In short, colonisation and (trans)plantation together with cultural dispossession and capital accumulation were not enterprises that European states, corporations and Churches pursued only once they were formed. They were formed to do it. Neither did they pursue these developments in closely integrated and uniform ways. The processes were diverse and changing, they warred with one another as much as they did over the objects of their material and spiritual ambitions (Hont 2005; Braudel 1982; Tilly 1990; Cook 2007). Christian missions, plantations and colonies were similarly also as reliant upon slavery (Stange 1968) as they were involved in relentless patronage-seeking and commerce, in ways in which Francke excelled in his day and evangelical churches especially in the US excel in their hyper-mediated ways today.

If missionising is thus a differentiating practice in which you refine and define your rules of truth, imperialising is a dividing practice in which you refine and define your truths of rule. The colony was always already an invention of the metropolis, the metropolis always already an invention of the colony. The two served to reflect and refract one another. There was no original. In pursuing their mutually disclosive belonging together, even as the American colonies did in revolt against the British Empire, metropolis and colony emerged through differentiating practices that often also bound imperialists, revolutionaries and the colonised together, in ways more enduring than any of them desired or cared to admit.

In consequence, and however much it was a source of controversy and dispute, imperialising was never a political after-thought of the religio-political nexus of the modern age. Such disputes as it engendered were constitutive of the way conflict over the rule of imperial truth shaped the truth of imperial rule, and thus of modernisation and development itself. The same obtains in respect of the modernisation and development programmes of the last twenty-five years.

However much it, too, was a source of controversy and dispute, neither was missionising a Christian after-thought. Such disputes as it engendered were constitutive also of how missionising Christian truth shaped the truth of missionising Christian rule. Consequently, you did not join the missions once you believed. You fashioned your faith in the missionising that you did. You did not imperialise after having formed your modern institutions and practices of rule. You fashioned your modern practices of rule through the very imperialising that you did. What was true in the seventeenth century remains true today.

Regina Schwartz made some remarkable reflections on sacramental poetics and the transubstantiation of Christian transubstantiation into the modern age in a way that provides a fitting conclusion here. Observing that, now living in an age of conflicting identities, 'each asserting their particularity against another, the result is invariably violent', she nonetheless also recognised that the 'opposite demand for a universal is attended by another kind of violence, the risks of political totalitarianism, a global imperialism, a violence that crushes particularity in its relentless drive towards universal control' (Schwartz 2008: 140). However much it has worked, most recently, through the politics of subjectification, particularities of ethnicity, religion, the specified freedoms of the market and the biopolitical manipulation of populations, in the pursuit of modernisation, security and developmental neo-liberal governance, preaching these as universals, engenders much the same outcomes.

This millennial-old problematisation of the rule of Christian truth and the truth of Christian rule did not, therefore, produce a solution to a problem. It instituted a complex and still mutating field of problematisation, formation and intervention driven by very many surfaces of friction and violently unresolved controversies of both truth and rule. The arcane dispute over the appointment of female bishops within the Anglican Communion is a contemporary case in point. Victoria Kahn's account of the English Revolution of the seventeenth century beautifully illustrates also how theology and religious discourse simultaneously worked as political theory and political discourse in ways that served to undermine and revise the truths of rule and rules of truth of medieval and scholastic traditions in the process of producing new ones (Kahn 2004; Kahn *et al.* 2006). To cite Regina Schwartz, 'If the Reformation inaugurated the beginning of modernism – the death of God and the birth of the modern Self', introducing processes that newly enslaved as they emancipated, 'it also transfigured values that in the earlier regime and in another register were conveyed as God-given: justice, love, and a living universe' (Schwartz 2008: 139). New truths of rule and rules of truth emerged to give concrete political and governmental form, including novel expressions of justice and love as well as novel demands for their fulfilment, to the outright rejection as well as revision of Christian truth and rule.

Correlations of truth and rule thus constitute fields of formation, intervention and application which give rise to changing idiomatic (specific, historically contingent, spatio-temporally located) problematisations and strategic formations of politics, government and rule. These problematisations necessarily also presuppose changing problematisations of space and time, of epochs of time and periodisations of history as well as of locations, both virtual and territorial, in which they take place, in addition to the technologies by which they are operationalised. What times are these? In what place or places do we find ourselves? What is the truth of the matter? Such questions help determine the problematisation of politics, government and rule. For time, place and truth are not simply containers. The specification of time and place are categories of truth, simultaneously in fact of both truth and rule, to which rules of truth and truths of rule give concrete form. Their very specification is designed by the rules of truth and truths of rule to make demands upon us, shape our conduct and circumscribe the imaginaries in which we dream about what we are, who we are, and what we can and must do.

Humans also come and go, as do all their rules of truth and truths of rule. Both are finite (Stambaugh 1992; Schmidt 1988; Hyland 1995; O'Byrne 2010). That finitude seems all the more marked because it is not only humans that come and go, it is their gods as well. From Nietzsche's death of God to Heidegger's keening for the presence of a god, the flight of the pagan gods is complemented by the withdrawal of the Christian God (Bulhof and ten Kate 2000; Schwartz 2008). God and gods leave the world repeatedly, and repeatedly return often in more violently worshiped forms. Eschatological sensibility both secular and religious is a function of finitude. Thus, the obsessive security–development politics of the twenty-first century are a secular, Judeo-Christian eschatology, a spiritual expression of it, for our current politics of security and development are as saturated as revealed religion with the fear of the end and the organisation of government and rule around it.

There are, however, key differences. For example, the eschaton of revealed religion marks a rupture between the immanent and the transcendent. The eschaton of modern finitude is immanent alone. It continuously punctuates the infinity of finite things of which modern finitude is comprised (Dillon 2011a). For each, the eschaton is as much a source of violent division as it is a foundational expression of theologico-political faith, a generative principle of formation for rule and misrule alike (Toscano 2010; Fiddes 2000; Ratzinger 1988; Schwarz 2000). In at least one very important respect, however, modern secular eschatology is more unforgiving than its religious counterpart. Its eschatologically driven security–development complexes offer no end to their violently modernising emergency politics of the end.

Notes

1 The *decorumstreit* in Halle - a conflict over the reconstruction of the conduct of conduct up to and including the very semantics and grammar of gesture - was long preceded by that which the advent of baroque politics witnessed in every domain of human conduct: from representation (painting) and science (optics, observation and experiment) to stagecraft, government and statecraft (Maravall 1968; and Jonckheere 2012). Via diverse idioms this took place from the Catholic south of Europe (during the correlation of baraoque politics with the Counter-Reformation) to the Protestant north (correlated albeit differently there with evangelicals as well, for all their piety and modesty of taste). We might add the additional reflection that decorum arises in Foucault, at least in part, through 'askesis' (McGushin 2007).

Bibliography

Afary, Janet and Anderson, Kevin (2005) *Foucault and the Iranian Revolution: Gender and the Seductions of Islamism* (Chicago, IL: University of Chicago Press).

Agamben, Giorgio (2011) *The Kingdom and the Glory: For a Theological Genealogy of Economy and Government* (*Homo Sacer* II, 2) (Stanford, CA: Stanford University Press).

Anderson, Benedict (1991) *Imagined Communities: Reflections on the Origin and Spread of Nationalism* (London and New York: Verso).

Auffarth, Christoph and Stuckenbruck, Loren T. (2004) *The Fall of the Angels* (Leiden: Brill).

Augustine, Saint (1984) *Concerning the City of God against the Pagans*, ed. Henry Scowcroft (Harmondsworth: Penguin).

Barry, Andrew, Osborne, Thomas and Rose, Nikolas S. (eds) (1996) *Foucault and Political Reason: Liberalism, Neo-Liberalism, and Rationalities of Government* (Chicago, IL: University of Chicago Press).

Benjamin, Walter (1998) *The Origin of German Tragic Drama* (London and New York: Verso).

Bernasconi, Robert (1998) 'Different Styles of Eschatology: Derrida's Take on Levinas' Political Messianism', *Research in Phenomenology*, 28(1): 3–19.

Blumenberg, Hans (1985) *The Legitimacy of the Modern Age* (Cambridge, MA: MIT Press.).

Brantlinger, Patrick (1996) *Fictions of State: Culture and Credit in Britain, 1694–1994* (Ithaca, NY: Cornell University Press).

Braudel, Fernand (1982) *Civilization and Capitalism: 15th–18th Century*, 2 vols (New York: Harper & Row), vol. 2.

Brewer, John (1988) *The Sinews of Power: War, Money, and the English State, 1688–1783* (Cambridge, MA: Harvard University Press).

Buck, Peter (1977) 'Seventeenth-Century Political Arithmetic: Civil Strife and Vital Statistics', *Isis*, 68: 67–84.

Buck, Peter (1982) 'People Who Counted: Political Arithmetic in the Eighteenth Century', *Isis*, 73: 28–45.

Bulhof, Ilse Nina and ten Kate, Laurens (eds) (2000) *Flight of the Gods: Philosophical Perspectives on Negative Theology* (Kampen: Kok Agora).

Butler, Judith (1997) *The Psychic Life of Power: Theories in Subjection* (Stanford, CA: Stanford University Press).

Cassedy, James H. (1969) *Demography in Early America: Beginnings of the Statistical Mind, 1600–1800* (Cambridge, MA: Harvard University Press).

Certeau, Michel de (1992) *The Mystic Fable* (Chicago, IL: University of Chicago Press).

Chamayou, Grégoire (2012) *Manhunts: A Philosophical History* (Princeton, NJ: Princeton University Press).

Chan, Simon (2011) *Pentecostal Theology and the Christian Spiritual Tradition* (Eugene, OR: Wipf and Stock).

Clarke, Thomas (2007) *International Corporate Governance: A Comparative Approach* (London and New York: Routledge).

Cook, Harold John (2007) *Matters of Exchange: Commerce, Medicine, and Science in the Dutch Golden Age* (New Haven, CT: Yale University Press).

Cooper, Melinda and Waldby, Catherine (2014) *Clinical Labor: Tissue Donors and Research Subjects in the Global Economy* (Durham, NC: Duke University Press).

Daston, Lorraine and Stolleis, Michael (eds) (2008) *Natural Law and Laws of Nature in Early Modern Europe: Jurisprudence, Theology, Moral and Natural Philosophy* (Farnham: Ashgate).

Davis, Kathleen (2008) *Periodization and Sovereignty: How Ideas of Feudalism and Secularization Govern the Politics of Time* (Philadelphia, PA: University of Pennsylvania Press).

Davis, Mike (2002) *Late Victorian Holocausts* (London: Verso).

De Goede, Marieke (2005) *Virtue, Fortune and Faith: A Genealogy of Finance* (Minneapolis, MN: University of Minnesota Press).

Dean, M. (2009) *Governmentality: Power and Rule in Modern Society* (London: Sage).

Debord, Guy (1994) *The Society of the Spectacle* (New York: Zone Books).

Dietrich, John W. (2007) 'The Politics of PEPFAR: The President's Emergency Plan for AIDS Relief', *Ethics and International Affairs*, 21(3): 277–92.

Dillon, Michael (2008) 'Security, Race and War', in Michael Dillon and Andrew W. Neal (eds), *Foucault on Politics, Security and War* (Basingstoke: Palgrave Macmillan), pp. 166–96.

Dillon, Michael (2011a) 'Acontecimentalidad: politica de la verdad y analitica de la finitud', in MiguelVatter and Miguel Ruiz Stull (eds), *Política y acontecimiento* (Mexico City: Fondo de Cultura Económica), pp. 271–311.

Dillon, Michael (2011b) 'Specters of Biopolitics: Finitude, Eschaton, and Katechon', *South Atlantic Quarterly*, 110(3): 780–92.

Dillon, Michael and Reid, Julian (2009) *The Liberal Way of War: Killing to Make Life* (Abingdon: Routledge).

Dubber, Markus Dirk and Valverde, Mariana (eds) (2006) *The New Police Science: The Police Power in Domestic and International Governance* (Stanford, CA: Stanford University Press).

Duffield, Mark R. (2001) *Global Governance and the New Wars: The Merging of Development and Security* (New York: Zed Books).

Duffield, Mark R. (2007) *Development, Security and War* (Cambridge: Polity Press).

Egginton, William (2010) *The Theater of Truth: The Ideology of (Neo)Baroque Aesthetics* (Stanford, CA: Stanford University Press).

Erdman, Carl (1977) *The Origin of the Idea of Crusade* (Princeton, NJ: Princeton University Press).

Eze, Emmanuel Chukwudi (ed.) (1997) *Race and the Enlightenment: A Reader* (Cambridge, MA: Blackwell).

Fiddes, Paul S. (2000) *The Promised End: Eschatology in Theology and Literature* (Oxford and Malden, MA: Blackwell).

Forsyth, Neil (1987) *The Old Enemy: Satan and the Combat Myth* (Princeton, NJ: Princeton University Press).

Foucault, Michel (2003) *Society Must Be Defended: Lectures at the Collège de France, 1975–76* (London: Picador).

Foucault, Michel (2007) *Security, Territory, Population: Lectures at the Collège de France, 1977–78*, ed. Michel Senellart (Basingstoke: Palgrave Macmillan).

Gal, Offer and Chen-Morris, Raz (2013a) *Baroque Science* (Chicago: University of Chicago Press).

Gal, Offer and Chen-Morris, Raz (eds) (2013b) *Science in the Age of the Baroque* (Dordrecht: Springer).

Gerhardt, Hannes (2010) 'The Problematic Synergy Between Evangelicals and the US State in Africa', in Jason Dittmer and Tristan Sturm (eds), *Mapping the End of Times: American Evangelical Geopolitics and Apocalyptic Visions* (Farnham: Ashgate), pp. 157–82.

Glete, Jan (2002) *War and the State in Early Modern Europe: Spain, the Dutch Republic, and Sweden as Fiscal-Military States, 1500–1660* (London and New York: Routledge).

Grundmann, Herbert (2005) *Religious Movements in the Middle Ages* (Notre Dame, IN: Notre Dame University Press).

Harris, Joseph (2014) *Inventing the Spectator: Subjectivity and the Theatrical Experience in Early Modern France* (Oxford: Oxford University Press).

Hearn, Julie (2002) 'The Invisible NGO: Evangelical Missions in Kenya', *Journal of Religion in Africa*, 32(1): 32–60.

Hont, Istvan (2005) *Jealousy of Trade: International Competition and the Nation State in Historical Perspective* (Cambridge, MA: Belknap Press).

Hunter, Ian (2006) *Rival Enlightenments: Civil and Metaphysical Philosophy in Early Modern Europe* (Cambridge: Cambridge University Press).

Hunter, Ian (2007) *The Secularisation of the Confessional State: The Political Thought of Christian Thomasius* (Cambridge: Cambridge University Press).

Hunter, Ian, Ahnert, Thomas and Grunert, Frank (eds) (2007) *Essays on Church, State, and Politics* (Indianapolis: Liberty Fund).

Hyland, Drew A. (1995) *Finitude and Transcendence in the Platonic Dialogues* (Albany: State University of New York Press).

Ingram, Alan (2010) 'Government and Security in the President's Emergency Plan for AIDS Relief (PEPFAR)', *Geogorm*, 41: 607–16.

Johnson, Adrian L. (2014) *War in Peace: British Military Operations since 1991* (London: RUSI).

Jonckheere, Konraad (2012) *Antwerp Art After Iconoclasm: Experiments in Decorum 1566–1585* (New Haven, CT: Yale University Press).

Joyce, Patrick (2003) *The Rule of Freedom: Liberalism and the Modern City* (London and New York: Verso).

Kahn, Victoria Ann (2004) *Wayward Contracts: The Crisis of Political Obligation in England, 1640–1674* (Princeton, NJ: Princeton University Press).

Kahn, Victoria Ann, Saccamano, Neil and Coli, Daniela (eds) (2006) *Politics and the Passions, 1500–1850* (Princeton, NJ: Princeton University Press).

Kantorowicz, Ernst Hartwig (1997) *The King's Two Bodies: A Study in Mediaeval Political Theology* (Princeton, NJ: Princeton University Press).

Keck, David (1998) *Angels and Angelology in the Middle Ages* (Oxford: Oxford University Press).

Lambert, Gregg (2008) *On the (New) Baroque* (Aurora, CO: Davies Group).

Lubac, Henri de (2006) *Corpus Mysticum: The Eucharist and the Church in the Middle Ages – Historical Survey* (London: SCM Press).

McGushin, Edward (2007) *Foucault's Askesis: An Introduction into the Philosophical Life* (Evanston, IL: Northwestern University Press).

Maravall, José Antonio (1986) *Culture of the Baroque: Analysis of a Historical Structure*, trans. Terry Cochran (Minneapolis: University of Minnesota Press).

Markus, R. A. (1988) *Saeculum: History and Society in the Theology of St Augustine* (Cambridge: Cambridge University Press).

Marshall, Ruth (2010) 'The Sovereignty of Miracles: Pentecostal Political Theology in Nigeria 1', *Constellations*, 17(2): 197–223.

O'Byrne, Anne E. (2010) *Natality and Finitude* (Bloomington, IN: Indiana University Press).

Osinga, Frans P. B (2007) *Science, Strategy and War: The Strategic Theory of John Boyd* (London and New York: Routledge).

Pagden, Anthony (1993) *European Encounters with the New World* (New Haven, CT: Yale University Press).

Parker, Geoffrey (1988) *The Military Revolution: Military Innovation and the Rise of the West, 1500–1800* (Cambridge: Cambridge University Press).

Patterson, Amy S. (2011) *The Church and AIDS in Africa: The Politics of Ambiguity* (Boulder, CO: First Forum Press).

Peterson, Erik (1964) *The Angels and the Liturgy* (New York: Herder and Herder).

Petty, William (1690) *Political Arithmetick* (London: Printed for Robert Clavel at the Peacock, and Hen. Mortlock at the Phoenix in St. Paul's Church-yard).

Piketty, Thomas (2014) *Capital in the Twenty-First Century* (Cambridge, MA: Harvard University Press).

Pocock, J. G. A. (2003) *The Machiavellian Moment: Florentine Political Thought and the Atlantic Republican Tradition* (Princeton, NJ: Princeton University Press).

Ratzinger, Joseph (1988) *Eschatology, Death and Eternal Life* (Washington, DC: Catholic University of America Press).

Reed, Annette Yoshiko (2005) *Fallen Angels and the History of Judaism and Christianity: The Reception of Enochic Literature* (Cambridge: Cambridge University Press).

Robbins, Joel (2010) 'Anthropology, Pentecostalism, and the New Paul: Conversion, Event, and Social Transformation', *South Atlantic Quarterly*, 109(4): 633–52.

Rusnock, Andrea Alice (2002) *Vital Accounts: Quantifying Health and Population in Eighteenth-Century England and France* (Cambridge: Cambridge University Press).

Schmidt, Dennis J. (1988) *The Ubiquity of the Finite: Hegel, Heidegger, and the Entitlements of Philosophy* (Cambridge, MA: MIT Press).

Schmitt, Carl (2005) *Political Theology: Four Chapters on the Concept of Sovereignty* (Chicago, IL: University of Chicago Press).

Schwartz, Regina M. (2008) *Sacramental Poetics at the Dawn of Secularism: When God Left the World* (Stanford, CA: Stanford University Press).

Schwarz, Hans (2000) *Eschatology* (Grand Rapids, MI: W. B. Eerdmans).

Skinner, Quentin (1998) *Liberty before Liberalism* (Cambridge: Cambridge University Press).

Skinner, Quentin (2008) *Hobbes and Republican Liberty* (Cambridge: Cambridge University Press).

Stambaugh, Joan (1992) *The Finitude of Being* (Albany: State University of New York Press).

Stange, Douglas C. (1968) '"A Compassionate Mother to Her Poor Negro Slaves": The Lutheran Church and Negro Slavery in Early America', *Phylon*, 29(3): 272–81.

Stoler, Ann Laura (1995) *Race and the Education of Desire: Foucault's History of Sexuality and the Colonial Order of Things* (Durham, NC: Duke University Press).

Tellenbach, Gerd (1948) *Church, State and Christian Society at the Time of the Investiture Contest* (Oxford: Basil Blackwell, ACLS Humanities E-Book).

Tilly, Charles (1990) *Coercion, Capital, and European states, AD 990–1990* (Cambridge, MA: Blackwell).

Todorov, Tzvetan (1999) *The Conquest of America: The Question of the Other* (Norman: University of Oklahoma Press).

Toscano, Alberto (2010) *Fanaticism: On the Uses of an Idea* (London and New York: Verso).

Walker, M. (2002a) 'Discipline and Decorum: The Decorumstreit' (unpublished paper).

Walker, M. (2002b) 'The Halle Enlightenment: Discipline and Decorum in Prussia, 1696–1702' (seminar paper, Department of History, Johns Hopkins University).

Watt, Nicholas (2010) 'Protests as UK security put at heart of government's aid policy', *The Guardian*, 30 August.

Whitmer, Kelly Joan (2008) 'Learning to See in the Pietist Orphanage: Geometry, Philanthropy and the Science of Perfection, 1695–1730' (University of British Columbia).

Wilson, Kalpana (2012) *Race, Racism and Development: Interrogating History, Discourse and Practice* (New York: Zed Books).

Wright, Stephen C. (2009) 'The Next Generation of Collective Action Research', *Journal of Social Issues*, 65(4): 859–79.

6

BIOPOLITICS OF SECURITY, RACE AND WAR

> Wars are no longer waged in the name of a sovereign who must be defended; they are waged on behalf of the existence of everyone; entire populations are mobilized for the purposes of wholesale slaughter in the name of life necessity.
>
> (Foucault 1998: 137)

Introduction

Race first figures as central to Foucault's analytic of modern power and politics during the course of the series of lectures that he gave in the mid-1970s; specifically in *Society Must Be Defended* (2003), the first lecture course in the series. That analytic of power preoccupied the work of the 1970s. The lectures coincide, for example, with the publication of *Surveiller et Punir* (1975) and *La Volonté de Savoir* (1976).[1] Stuart Elden has also noted that they were contemporaneous in addition with the publication of Volume 1 of *The History of Sexuality*. *Society Must Be Defended* ran from January to March while *The History of Sexuality* appeared in December of 1976. As Elden concludes, 'the published volume being an overview of the projected six part series ... Volume 1 was in part a summary of lectures [*Society Must Be Defended*] that must surely have been written by the time the book was delivered to Gallimard' (Elden 2002: 127n). Some of the key analytical themes broached in 1976 are also pursued throughout the succeeding lecture courses *Security, Territory, Population*, 1977–8 (2007) and *The Birth of Biopolitics*, 1978–9 (2008).

These lectures describe a trajectory rather than a system of thought. It is a trajectory of thought concerned with broaching a novel analytic of modern politics, specifically liberal biopolitics. Admittedly an incomplete project, Foucault is in effect rewriting the history of the birth of modern liberal politics in a genealogy which typically focuses on the micro-practices of rule rather than the grand self-serving narratives of modernisation, liberation and enlightenment. He is, therefore, less interested in the ontologising manoeuvres of liberal political philosophy than he is in the historically evolving material practices of liberal forms of rule. However much it calls liberal political thought into question, his is an analytic of modern liberal power relations not an ideology critique of liberal political philosophy.

From the birth of political arithmetic in the seventeenth century through the avalanche of printed numbers in the eighteenth and nineteenth centuries through to the 'discipline' and 'control' societies of the twentieth century, biopolitical governance has sorted life into racially inscribed categories because that is what the operational logic of biopolitics obliged it to do. And while this racial inscription was most marked in the colonial despoliation of non-Western peoples, it was never pursued independently of the biopolitical colonisation of the metropolitan populations of the Atlantic world. It began there, for example in the population surveys which accompanied the early seventeenth-century colonial plantations in Ireland.[2]

Foucault tells us that the object of biopolitics is to make life live. But it is equally clear that, when one follows his analytic carefully, especially in *Society Must Be Defended* (2003), and that when one extrapolates the logic of biopolitics as well, then making life live is evidently a lethal business which makes war on life which does not fit the template of biopoliticised life and its ways of making life live. Making live is simultaneously also a making die for the sake of Life.

Where life is improvable, biopolitics specifies continuous revision and reform. Where life is, however, obdurately resistant to biopolitical revision, biopolitics specifies correction and punishment.[3] Where life simply exceeds biopolitical rationalisation and technological governance – wherever life proves itself biopolitically ungovernable, unclassifiable or incalculable – biopolitics terrorises life and, in many varied ways, specifies death (Dillon 2007a). Since it is a condition of making life live that biopolitics inevitably encounters life inimical to its project as an everyday fact of biopolitical life, the peace which biopolitics seeks 'makes live' through continuous warring against life which does not fit; especially against life which endangers life's biopoliticisation. Biopolitically, that which is not fit becomes unfit. Biopolitically, that which is not meet becomes meat. In order to differentiate between meet and meat, biopolitics is required to assay and grade life. In short, not all life, not every conceivable kind of life, is biopolitically suitable life. It always turns out that in biopolitics some life has to die in order for some other life to live. In that sense biopolitics is simultaneously also a politics of life and death. What differentiates biopolitics from sovereign politics is a change in the correlation of life and death, not some escape from the inevitability of that correlation.

Race is one of the markers which biopolitically adjudicates Life. It does not only specify life's eligibilities – for this or that good – it ultimately specifies whether or not a life is to be considered eligible for life as such. Political enfranchisement of life, biopolitically, is ultimately dependent upon utility measures for the promotion of life biologically. The individual and the species are intimately tied in to one another governmentally.

This biopolitical war for Life against life is, therefore, no accident. It is not epiphenomenal, and it has no obvious end. For life must continuously exceed the biopolitical speciation to which biopolitics must reduce it, if biopolitics is to make life live in the ways in which biopolitics understands Life to be. In short, the problem which biopolitics encounters in pursuit of its project to 'make life live' is life itself. The war for life which biopolitics wages on behalf of its understanding of Life and in relentless pursuit of appropriate power relations to enact that understanding, is translated into biopolitical peace through an obsession with security. Biopolitically it is Life which has to be secured against life. Peace is written as war biopolitically through discourses of security. To make life live it has to be secured. Securing life is a continuous war against whatever threatens life. Life is thus a permanent security problem for biopolitics.

The reason is very simple. Life is an undecidable, not a datum. It comes indebted to the gift of life which is more than biopolitical Life itself. It also comes in many more shapes and sizes, many more expressions, desires, tones, affectivities, aesthetics and so on, than that presupposed by biopolitics, and it constantly changes in ways which call down the violent opprobrium of biopolitics. So, if we are to fully appreciate the analytical drivers and political dynamics of biopolitics, and if we are to understand its war-making through constant securitising, it is not enough to assume that we have exhausted what Foucault opened up to interrogation when we follow his saying that biopolitics is committed to making life live.

One has to note several other things as well. First, that Life, here, is the biopoliticised reduction of life. Second, that the biopoliticised reduction of life is not an accomplishment. It is a contested and contestable form of government and rule in which concrete political form is given to existence enframed by factical finitude. Third, the psychic life of power revolves, here, around the cultivation of a voluntary servitude to the infinity of finite governance in which life is held continually on probation. On probation for what? For the test of its continuing governability and fitness for continually changing forms of self-governance that now count as freedom. I call this its *mentalité du gouvernement*. Fourth, one has to say that this biopolitical project of making life live is also a lethal business. Life cannot be made, encouraged, or otherwise seduced, into living biopolitically unless other ways of living and other forms of life are extinguished.

Foucault recognised that the lethal politics of making life live must also refigure death as it refigures life. In response to its biopoliticisation, he said: 'Death was no longer something that suddenly swooped down on life ... Death was something permanent, something that slips into life, perpetually gnaws at it, diminishes it and weakens it' (Foucault 2007: 244). He saw that death was not expunged from life, but that life and death were newly linked in biopolitics. In this emergent correlation of life and death, he saw that biopolitics newly problematised death as much as it did life:

> I wouldn't say exactly that sovereignty's old right – to take life or let live – was replaced, but it came to be complemented by a new right, which does not

> erase the old right but which does penetrate it, permeate it. This the right, or rather precisely the opposite right. It is the power to 'make' live and 'let' die.
> (Foucault 2007: 241)

He briefly experimented with calling this refiguration of death 'thanatopolitics'.

Biopolitics does not, however, simply refigure death in the process of refiguring life. It begins to establish a new economy of life and death. In relation to how biopolitics newly instrumentalises killing and death, for example, Foucault observed: 'When I say killing I obviously do not mean simply murder as such, but also every form of indirect murder: the fact of exposing someone to death, increasing the risk of death for some people, or quite simply political death, expulsion, rejection, and so on' (Foucault 2007: 256). Thus, as Achille Mbembe and others following Foucault have noted, since you cannot make life live biopolitically unless you simultaneously also 'let die', biopolitics does not simply offer a thanato-political refiguration of death. It does that and more. Its newly emergent political economy of life and death systematically installs what Mbembe has called a necropolitics of dead life. The life of the living dead now characterises our global biopolitical economy quite as much as the living life of biopolitics (Mbembe 2001, 2003; Montag 2005). Hence, in my conclusion, I pose the characterisation of 'Man' and Life as dead and deadening strategic figures, in whose after-lives we remain governmentally enthralled.

Necropolitics is the 'letting die' required by the biopolitical injunction to 'make live'. Making live must systematically adjudicate the living in respect of their contribution or otherwise to the project of making Life live. Just as making live can be a violent process, so letting die is no mere accident. It is here that racial markers are commonly employed to operationalise the biopolitical adjudication required to follow the injunction to make life live. Hence biopolitics, necropolitics and race are closely correlated phenomena. Race contributes directly to the triangulation of biopolitics with its necropolitics. That triangulation helps strip biopolitics of any assumed innocence in respect of its project of making life live.

From Foucault's perspective, therefore, the story of biopoliticised race arises with the birth of political modernity. It is part of that process by which the conduct of conduct comes to be governed by other mechanisms than those once provided by Christian Kingship and the pastoral power of the Christian Church. But one has to ask: what is this Life which liberal biopolitics takes as its referent object of power and politics? It is not the life of Aristotle's *De Anima*, or the *vivere civile* of Renaissance humanism famed by Machiavelli. It is not identical either with Descartes *cogito*, or with the life of possessive individualism said to distinguish the normative political philosophy of liberal political theory.

Life here is, instead, the biological account of life which first arises during the course of the seventeenth and eighteenth centuries, especially in the form of 'population'. It develops, thereafter, through accounts of life furnished by eighteenth-century thinkers and proceeds on into the nineteenth century when biology was born and the term Life itself first became a scientific term of art (Jacob 1989). It develops through the Darwinism of the nineteenth and Neo-Darwinism of the twentieth century and, thence, into the molecularisation of life and the dramatic proliferation of the life sciences which now characterise the early twenty-first century. In short, the life of biopolitics has a history, and that history tracks changing racial as well as biological accounts of what it is to be a living thing. The history of the one was always closely associated, and even sometimes fully conflated, with the other. If racism has its history, so also does the biopoliticised understanding of life as species existence. The two have always been closely entwined.

The biopolitical genealogy of modern power and politics

> … at stake is the biological existence of a population.
>
> (Foucault 1998: 137)

Following the religious revolutions of the seventeenth century, especially, politics and rule were obliged to seek some other warrant than that of a divinely ordained cosmos. However much they nonetheless continued to rely on some guarantor of universal order, that guarantor could no longer be confidently formulated in terms of a providential God divinely manifested throughout creation. Science, commerce, urbanisation, religious war and the rise of the state, as well as the bloody rivalry of conflicting confessions of faith, put paid to that game. What, therefore, first distinguished modern accounts of government and politics, for Foucault, was the way in which they sought, in the problematisation of all accounts of transcendent order, to ground the problematic of rule, instead, in some internal, or immanent, political rationality. In the process of documenting how the biopolitical aspects of this historical transformation of politics, government and governance emerged, Foucault makes a whole variety of arguments. I boil these down to emphasise the following four points.

First, Foucault maintains historically that political modernity arose from war, specifically race war. Second, he observes that early biopolitics, taking life as its referent object of power and politics, focused on the empirical referent of 'population'. Third, he extended his initial argument by maintaining that not only did political modernity arise out of race war, but he insisted that the very grid of intelligibility of modern politics itself continues to derive from war. I describe this as a process in which the *logos* of war is inscribed as the *logos* of peace. Arguing with Foucault, but beyond Foucault, I seek to derive a fourth argument. I maintain, in addition, that the *logos* of peace is systematically inscribed with the *logos* of war through discourses of security. What is enunciated at the level of ontopolitical necessity – the reality that species existence revolves around survival and that that life must first be secured if it is to be promoted – is also inscribed at the level of rule through the operation of biopolitical governmentality.

To the degree that contemporary politics is biopolitical, therefore, biopoliticised racism does not so much take place via political rationalities which proclaim the 'realism' of racial supremacy. It more regularly occurs instead, now, through the socio-technical systems which comprise the governing technologies of biopolitics. It is these which more regularly exercise that everyday discrimination against lives required, biopolitically, to make life live. Within Atlantic societies, these days, race and biopolitics meet – occur and recur – throughout the circulatory relays and institutions of all its governmental capillaries of power, rather than in officially endorsed political rationalities of race. A complex assemblage of power relations, or, what Foucault calls *dispositifs de sécurité*, racial markers help adjudicate the living for biopolitics (Foucault 2007).

Race war (La guerre des races)

> ... the blood that has dried on the codes of law.
>
> (Foucault 1986)

The modern politics which first arose in France and England during the course of the seventeenth century arose, according to Foucault, through war, and the wars through which they arose were race wars (*la guerre des races*). Here, Foucault contests the history and interpretation of the origin of early modern politics. His protagonists are many. They include, in particular, Machiavelli and Hobbes as well as Hegel. Machiavelli and Hobbes are challenged because the former is preoccupied with the politics of seizing and retaining a sovereign's grip on his territorial state (Machiavelli), and because the latter's account of politics derives from the posited abstraction of a war of all against all rather than real conflict, and the mortgaging of the political subject's life to the death dealing monopoly of the sovereign (Hobbes). Hegel's dialectic is challenged because Foucault does not subscribe to the idea of history having a goal or *telos* and he does not believe in the idea of a historical subject coming to fruition through history. He was never that kind of historian. Hegel is also implicitly contested because, like Hobbes, he also substitutes an ontological mechanism for the historical fields of force which were instituted by real wars: 'invasions, struggles, plundering, disguises, ploys' (Foucault 1986: 76; see also Marks 2000). Since Foucault describes how these early modern wars were race wars, race figured from the very beginning as one of the ensemble of dividing practices which gave birth to modern discourses of the political, and the apparatus of modern economies and state-craft. He describes how, with the dissolution of the Christian world, civil war and emerging conflicts over the institution of absolute monarchy, jurists, political theorisers and popularisers, in both England and France, appealed to the traditions of their 'race' – Anglo-Saxon, or Frank, for example – to proclaim or protest newly emergent forms of rule.

In the late seventeenth century, and throughout the course of the eighteenth, liberal biopolitics began to find its warrant, instead, in the biological understanding of life as species being. It enacted that warrant, especially against the sovereign claims of absolute monarchy, by seeking to align government with the limits set by what were increasingly said to be the independent dynamics of life processes; such as those exhibited, for example, by 'population' and 'economy'. Good government then became government which not only respected the limits within which it should govern, but artfully sought to manipulate and manage those limits for the good governance of species life.

Life thus came to teach liberal biopolitics what it needed to know if it was to promote, regulate and govern life effectively according to the 'natural' limits set by the exigencies of species existence itself. Crucially, those limits, and the life processes which they demarcate, had to be amenable to calculation – to their own specific forms of power/knowledge. If they were not, then liberal biopolitics would quite literally not know what to govern, or how to govern. Where bodies and populations, nonetheless, appear to exceed or escape calculation in some way or other, however, there monstrosity lies for the biopolitical technician and for the biopolitician. This fear was regularly expressed in racial terms. It found expression in the early political arithmetic of seventeenth and eighteenth biopolitics, just as it finds expression today in the biometrics and risk calculations which enact twenty-first century biopolitics: especially in the allied campaigns of 'homeland security', 'national resilience' and the 'war on terror' as well as throughout immigration and asylum policies.

But one has to be clear about what Life comes to mean under this emergent biopolitical regime.

Species life

> '*le genre humain*' to '*espèce humaine*'.
>
> (Foucault 2007: 75)

The human can be classified in many ways; indeed the term 'human' is itself a form of classification. Foucault was interested in what happens to politics and power relations, and why, in particular, race becomes such an integral mechanism of power, when modern politics and power come to understand and operationalise themselves as regulating and promoting something called *espèce humaine* rather than *le genre humain*?

Recall, then, that intrigued by the mechanisms through which the basic biological features of the human species became the object of political strategy, when Foucault first came to interrogate biopolitics he noted how it differentiated life as species existence from life understood in other ways. Life understood as species existence – *espèce humaine* – differs fundamentally, he records, from life understood as '*le genre humain*' (Foucault 2007). As related earlier, the root of *le genre humain* – *gens* – refers to the *jus gentium* of Roman and medieval law.

The move from *gentium* to *espèce* thus effects a transformation in the very understanding of what it is to be a living being, and, correspondingly, of the governmental regulation of such an object. At the end of the seventeenth century, that object took the specific form of 'population'. By the twenty-first century, it has increasingly come to take the form of informationally constituted complex adaptive living assemblages. It is here that the biopolitics of security of the twenty-first century incorporates into the infinite government of finite things the infinite governance of beings-in-formation, or emergence, continuously becoming-dangerous to themselves and the contemporary rules of truths and truths of rule through which they are governmentally constituted: becoming-dangerous, indeed, via the very complex processes of adaptation and emergence said to constitute them. For that reason, we can now say that there is a very large extra-human, indeed 'post-human', component to contemporary biopolitics of security incorporating the living assemblages to be found in the domains, for example, of bio-security, cybernetic security and virtual security (Dillon 2003).[4]

It was first 'police' and then liberal biopolitics which took population as its referent object of power and security. They differed from each other in as much as 'police', or *polizeiwissenschaft*, was a device – a kind of theory of power in itself – deployed on behalf of the newly emerging modern sovereignties. It was responsible for the early introduction and refinement of population statistics, originally called 'political arithmetic', and it treated population as a resource for the newly empowered modern sovereign. The principal difference between police and liberal biopolitics was that the latter effectively democratised statistics, in the sense of embracing them for progressive social change rather than resisting them as an arm of sovereign power, attributed autonomy to the behaviour of populations and increasingly began to construe individual and collective freedom transactionally as the right to participate in markets.[5] Just as Foucault records how mercantilism constrained the operation of economy governmentally as well as commercially – the one becomes an operationalisation of the other – by limiting it to the sovereign, so also it constrained the field of application and regulatory potential becoming available through population and political arithmetic (early modern demographics, for example) by similarly limiting these to the horizons of possibility and operability constitutive of the rules of truth and truths of rule of sovereign power as one. As something to be called 'the economy' escaped its enframing by mercantilism and sovereignty, so also did political arithmetic and population. Population, particularly in the form of productivity and consumption, become integral to 'the economy' and especially also the market. Political arithmetic originating as a kind of early modern surveillance similarly also mutates during the course of the mid-eighteenth century – in the American colonies as much as in the European metropoles – into an instrument recording the transformation of the sources of wealth, aiding thereby the emancipation of the newly empowered sources of production and enrichment from the orders of class and privilege constitutive of early modern sovereign rules of truth and truths of rule.

Specifically, then, in relation to *espèce humaine*, power comes to be exercised on, in and through the biological mass which constitutes the 'species', rather than the juridico-political and cultural processes of belonging and rule said to constitute the *gens* of *gentium*, or of '*le genre humain*' (Bartelson 1995). Thus the principles of formation around which power relations and technologies of power began to emerge were governed by those features which most characterised species existence, rather than those said to characterise 'providence', 'nature' or, even, the 'history' and 'territoriality' of a 'people'.

Species existence is a survivalist existence which in our times is now set in a matrix of continuous emergent necessity subject to the radical contingency of events including, of course, the radical coincidences of time and space which comprise the events of its own existence, indeed comprise the radical contingency of its very own event. Here Foucault describes biopower as a technology,

> which is centred not upon the body but upon life; a technology which brings together the mass effects characteristic of a population, which tries to control the series of random events that can occur in a living mass, a technology which tries to predict the probability of those events (by modifying it, if necessary) or at least to compensate for their effects.
>
> (Foucault 2007: 249)

Foremost among the characteristic features of biopolitics noted by Foucault from its very inception, and amplified here for further analytical effect, looking forward to their critical importance in contemporary biopolitics, were therefore circulation, contingency and complex connectivity (Dillon 2005a, 2007). Circulation, because that was how the life form was said to propagate itself. Contingency, because its very existence was an aleatory phenomenon. Thrown into life, itself an 'accidental' phenomenon, species existence becomes the sum total of its encounters with the contingent conjunctures said to characterise the life of biological entities, human and non-human, in conjunction also with their environmental embeddedness. Complexity, because the trajectory of such existence was no mere projection of linear progress. If its time was evental, its experience of its own event in time was non-linear – comprised of complex adaptation, phase changes and unpredictable accidents and transformations.

What therefore followed in that shift from *gentium* to *espèce* was a transformation not only in the referent object of power relations, and not only also in the very mechanisms by means of which power operates and circulates. Above all, the temporality of such existence comes to be governed not by astrology, 'Providence', or even history but by 'evental time' here. I am invoking but simultaneously also taking some issue with Jacques Le Goff's fascinating account of the temporal transformations which took place in the middle ages, during the course of which he differentiates Church time from merchant time. Merchant time was not, however, simply linear as Goff maintains. It was also evental. But it was evental in ways that differed from the Church. Its evental time began to be infused, for example, by the logic and risk-based technologies of capital accumulation, rather than the Grace of God (Le Goff 1982).

Thus the biopolitical problematisation of security in terms of species existence is very different from that of the geopolitical problematisation of security in terms of territorial sovereignty, or the Christian soteriological problematisation of security in terms of salvation or redemption.

War as the logos of politics

> ... we must hear the distant roar of battle
>
> (Foucault 1979: 308)

A third key point to make about Foucault's formulation of the problematic of modern power and politics is that modern politics, founded historically in war, operates as an extension of war. War is not the extension of politics by other means, says Foucault. Quite the opposite, modern politics is an extension of war. Its very grid of intelligibility continues to extend the operational practices and discursive assumptions of the *logos* of war into the *logos* of peace. If race was essential to the early formation of the modern political imaginary in which liberal governance remains implicated so also was war.

Foucault is not saying, as many traditional theorists of political modernity have said, that war is the ontological foundation of political order. His argument is not, at least initially, an ontological, ontopolitical or polemological argument. It is an historical argument about how political modernity – the discursive practices and structures of the modern state and of liberal governance in particular – emerged historically.

This historical argument does have profound epistemological implications, however. If that was how political modernity and modern power relations emerged, then the analytic of modern power relations – how we come to know, interrogate and analyse the operation of modern power – ought then to revolve around its changing historical rules of truth and truths of rule, rather than metaphysical arguments concerning, for example, sovereign subjectivity, rights and nature.

Foucault describes what he calls this critical politics of truth in the inaugural lecture to a series of lectures he gave to the University of Louvain in 1981. There, Foucault distinguished between different types of critical philosophy in a summary that encapsulated the very way in which he worked as well as the things he worked upon. While his previous work had sought critical purchase upon the cruelties of apparatuses of truth, reason and rule, Foucault clearly also understood that the absence of transcendental guarantees for human judgement was not only the very condition of possibility for acquiring it. In the lectures he gave throughout the early 1980s he also explored how the acquisition of it was not a mere technique but a life's work.

He began with his inspiration in curiosity.[6] 'If critical philosophy is a philosophy that starts *not* from the wonderment that there is being, but from the surprise that there is truth, then we can clearly see that there are two forms of critical philosophy'[7] (2014: 20; emphasis added here, and in the following quotations). On the one hand he says, 'there is that which asks under what conditions – formal or transcendental – there can be true statements' (2014: 20). On the other, there is, 'that which investigates the forms of veridiction, the different forms of truth-telling' (2014: 20. He elaborates: 'In the case of a critical philosophy that investigates veridiction, the problem is that of knowing *not* under what conditions a statement is true, but rather what are the different games of truth and falsehood that are established, and according to what forms they are established' (2014: 20). And so: 'In the case of a critical philosophy of veridictions, the problem is *not* that of knowing how a subject in general may understand an object in general.' (The dividing of the world into subject and object provides the structural condition of possibility that Foucault calls *connaissance* which gives rise to 'truth' in the form of the positivities and empiricities of modern knowing.) The problem is instead, he says, 'that of knowing how subjects are effectively tied within and by the forms of veridiction in which they engage. In this case, the problem is not that of determining historical accidents, external circumstances, the mechanisms of illusions or ideologies, or even the internal economy of errors or failures in logic that could have produced the falsehoods' (2014: 20).

What then is the problem? 'The problem is to determine how a mode of veridiction, a *Wahrsagen* [Nietzsche], could appear in history and under what conditions' (2014: 20). And so: 'If, from the point of view of the true, history can only explain the existence or disappearance of falsehoods, from the point of view of veridiction, history can explain the appearance of a truth-telling' (2014: 20). 'Well you understand, no doubt,' he then says, 'that the object of a critical philosophy of veridictions is not to constitute a "general police" of the true, or to constitute a set of instruments that is general enough to fix the formal conditions according to which these statements could be true' (2014: 20). Rather this exercise, 'is a question of defining modes of veridiction in their plurality, of searching for the forms of obligation through which each mode binds the subject of truth-telling, of specifying the areas to which they apply and the fields of objects they bring to light and finally the relationships, connections and interferences that are established between them' (2014: 20). In sum: 'in this critical philosophy it is not a question of a general economy of the true, but rather of a historical politics, or a political history, of veridictions' (2014: 20). In short it is a matter of interrogating the inescapable co-implication of rules of truth and truths of rule.

Biopolitics of security as necropolitics

If Foucault is right, that liberal peace is the extension of war by other means, then those other means come in the form of the prevalence of security discourses. Peace becomes the extension of war through the discourse of security. Its very arts of governance revolve around the securing of life as species existence. Here is how and why.

Whatever endangers the promotion of species life endangers liberal biopolitics. The peace for which liberal biopolitics strives is that of the uninterrupted promotion of species existence. Such a peace is endangered when it is challenged by other accounts of existence and by the sheer intractability of species existence itself. Liberal biopolitics makes war on that which endangers species existence through the discursive practices which seeks to secure the promotion of species existence. Peace and war find their biopolitical articulation in the biopolitical discourses of security.

In sum, making life live ostensibly rejects war as a virtue and proclaims peace. The vocation of war is to kill. The vocation of biopolitics is 'to make live'. But biopolitics cannot make live unless it preserves life from that which threatens it, in order to promote the enhancement of life. To do that biopolitics must also seek a command of a refigured death, specifically that of biopoliticised economy of who shall live and who shall die. It is a dangerous game in every respect. Although he acknowledges it, Foucault does not reflect on this necropolitics beyond the observations he made about the state racism of Nazi Germany and the incipient racism of State Socialism. My additional argument is that it is apparatuses of security which, therefore, do the biopolitical work of inscribing the *logos* of peace with the *logos* of war. Liberal peace is a necropolitics of security which makes permanent war against life on behalf of life.

Making life live is therefore a lethal business because the promotion of species existence appears to be threatened on all sides not only by alternative accounts of existence but also by the danger which species existence always seems to pose to itself, not least in often being resistant to the biopolitical injunction to make life live. For not all life can live if life itself is to be promoted. Some life is inimical to life and has to be exterminated if it cannot be corrected and reformed. Life is like that. To be precise, species life is like that, and so we have to clarify this basic classification of what it is to be a living being because it is foundational to biopolitics and how, as such, it has need of the sub-division of species life into more or less functionally utile categories of human life to which the term race applies.

Foucault pursues the point through two well-formed questions: 'Given this power's objective is essentially to make live, how can it let die? How can the power of death, the function of death, be exercised in a political system centred upon biopower?' (Foucault 2003: 254). The answer is prefigured in how he habitually talks about power and politics in terms of political rationalities and governmental technologies. If governmental technologies regulate, political rationalities ontologise. They express an understanding of the real. In Foucault, ontologies matter, but he does not presume that material practices proceed unproblematically from ontological principles. The world is too messy for that. There is continual interplay between ontologising and technologising. An understanding of the real lurks in every technology. Every ontology desires to be operationalised in an appropriate technology.[8]

Biopolitics performs the ontologising and technologising through the simultaneous refiguration of both life and death. That is to say through the biopolitical enunciation of the real as species existence (political rationality), and its micro-political practices (governmental technologies). It is at this point, Foucault says, that 'racism intervenes' (Foucault 2003: 254). It breaks up the biological continuum, sub-divides the species, according to which forms of life are more fit, more eligible or more disposed to life and which are not; and which are indeed inimical to life and in need of extermination.

Here Foucault says, in addition, that racism enacts the relation of war which templates the modern account of the political: 'this relation ("If you want to live you must take lives, you must be able to kill") was not invented by either racism or the modern State. It is the relationship of war' (Foucault 2003: 255). Enacting the relation of war, the martial imprinting of modern politics in biological terms, biopolitics differentiates life into categories of living things more and less eligible to live by virtue of the ways in which they live, accounting whether or not that living promotes, diminishes or profoundly threatens species life itself: 'The fact that the other dies does mean simply that I live in the sense that his death guarantees my safety; the death of the other, the death of the bad race, of the inferior race (or the degenerate, or the abnormal) is something that will make life in general healthier: healthier and purer' (2003: 255). Biopolitics thus enacts a necropolitical audit of living things in which race functions as a sorting device. Racism is a kind of '*Appel*' in which those to be marked out for biopolitical discrimination, correction and, if necessary, elimination are distinguished and selected. This necropoliticised peace machine – which runs the gamut from peace-keeping and peace-making to 'operations other than war' and imperial conquest – is the liberal way of war (Dillon and Reid 2009).

In sum, Foucault is not concerned to give an account of race as such in these 1970s lectures. Race arises for him as a mechanism integral to the operational and analytic logics of a form of power and politics which is biopoliticised in as much as it takes human being in the form of species life as the referent object of power and politics. Not all modern politics does this, but an overriding and increasing proportion of it does.[9]

It is equally important to emphasise that, as with power so also with race, Foucault is not claiming that this biopolitical enframing of politics, power and race supersedes all other accounts of politics, power and race. Power is a diverse and changing manifold. So also is race. In politics as in liberal biopolitics these two are, however, intimately linked according to Foucault. Biopolitics requires race. Race is necessary to making biopolitics work:

> racism justifies the death function in the economy of biopower by appealing to the principle that the death of others makes one biologically stronger insofar as one is a member of a race or population, insofar as one is an element in a *unitary living plurality*.
>
> (Foucault 2003: 258)

Note the added emphasis. Foucault is analytically prescient. He recognises that it is not simply a matter of belonging to a kind of pre-established racial entity but the generic biologised notion of 'a unitary living plurality'. As doctrines and ideologies of racial supremacy lost their cultural and political purchase – their racialised accounts of the real failing to retain their previous ontopolitical grip on the cultural and political imaginaries of Atlantic societies – the logic of biopolitics nonetheless persists where life as belonging to a 'unitary living plurality' remains the generative principle of formation among them for governance and politics.

Therefore, just as in *The History of Sexuality* Foucault's problematic was not ultimately that of 'sex' but of the politics and power relations in which sex figures as a pivotal mechanism (Foucault 1998: 145),[10] so also in the lectures under discussion here the problematic is not ultimately that of race, colonisation or empire but the politics and power relations in which race figures as a point of application for power relations.[11] In each instance, sex and race figure as the points of application for the operation of distinctive forms of power and politics – their rules of truth and truths of rule. The same applies also to the freedom presupposed by biopolitics. It, too, is a point of application for technologies of self-governance and truth-telling practices which make that freedom objectify itself in self-regulating ways. Biopolitics is the background that therefore enables us to understand the importance assumed by race as a biopolitical device.

Early biopolitics: population, political arithmetic and race

> Such a power has to qualify, measure, appraise and hierarchize, rather than display itself in its murderous splendour.
>
> (Foucault 1998: 144)

Murderous splendor is as integral to neo-baroque politics as it was to the early baroque politics. But, we might hazard the thought that scopic, disciplinary and surveillance regimes, among many other things, change as you move from one to the other. Biopolitics therefore also changes. It must do so in as much as its referent object – the life of species existence – also changes. Biopolitics changes because species life also adapts. It adapts, for example, in response to biopolitical interventions designed to make species life change. Biopolitics also changes in response to changing scientific understanding of the nature of species existence and the growing capacity of digital and molecular science in particular now to intervene in life's very own creative – morphogenic – processes.

Two good examples are provided by population and biology. Within the Atlantic world especially, twenty-first century populations differ from eighteenth-century populations, and twenty-first century biology differs from nineteenth-century biology. In the first example, morbidity has been compressed. In the second instance, biology has been molecularised. The shape of population and the understanding of living processes are now radically different from what they once were. In the process, life's vital statistics have come to be taken differently. The biopolitical technologies and political rationalities by means of which life is governed, regulated and transformed have also changed. Biopolitics is thus a changing historical, and not a fixed, form of politics, government and rule. Its own rules of (biological) truth and truths of (biopolitical) rule change. They thus change as these and other changes occur. So also does the biopolitical operation of race in the enumeration of the vital statistics upon which the micro-practices of biopolitical governance are based.

Population

> ... the greatness and glory of a prince lyeth ... in the number, art, and industry of his people, well united and governed.
>
> (William Petty, 1662)

If population was the empirical referent for *espèce humaine*, the grid of intelligibility which surrounded the emergence of 'population' as a referent object of power and security in the modern period differed as much from sovereign accounts of power as it did from the pastoral power of the Church. It also bifurcated. The promotion of population in *polizeiwissenschaft* differed from the promotion of 'population' in liberal biopolitics. In the one it was a sovereign resource, in the other it became a political vocation. Before moving to the historical transformation of the vital biopolitical statistics concerning population which took place during the course of the eighteenth century – and of the changing role which race played in them – we have to clarify the significance of this shift in the referent object of power and politics to population.

For example, the concern with population did not derive from the Church's redemptive mission, determined as it was by the task of pursuing the salvation of everlasting life for its flock. In a sense it was the collapse of the Church's authority, and its allied structures of medieval Empire and Kingship, posing a generic problematic of the 'conduct of conduct' that helped give rise to power and politics taking population as the referent object of their political rationalities and governmental technologies. A population is, therefore, not something that is posited by, and regulated through, the intercession of the pastoral power of the Church mediating between man and God. It is not a soteriologically organised ecclesiastical community of faith in receipt of grace, shepherded by bishops and clergy through the law, the sacraments and the liturgy, bound in worship to an all-powerful providential creator.

Similarly, securing everlasting life for a divinely created existence is not the same kind of security project as that which aims to secure the promotion of life construed in terms of the biological materiality of species existence. The one is a soteriological mission which takes place within a divinely ordained cosmos. The other is a calculative challenge which takes place within a complex matrix of species existence governed by contingency, connectivity and complexity. Ultimately, as Atlantic societies became thoroughly capitalised and governed extensively through the market, this matrix (of necessary contingency as well as of contingent necessities) came to revolve around the commodification of contingency as risk, and the adoption of risk analysis as the dominant decision technology. However much it may be proclaimed as a political ontology, 'necessary contingency' is a political project revolving around a rigorously capitalised economy which increasingly conflates liberal freedoms with the multiplication of transactional market rights.

Similarly a population is not a polity. It is not a civil community in the Hobbesian or other early modern liberal tradition. In biopoliticised *dispositifs de sécurité* contra Hobbes, for example, Foucault explains that:

> the principle underlying the tactics of battle – that one has to be capable of killing in order to go on living – has become the principle that defines the strategy of states. But the existence in question is no longer the juridical existence of sovereignty; at stake is the biological existence of a population.
>
> (Foucault 1998: 137)

Thus, if the corollary of the Hobbesian account of politics, power and war was the acceptance of Leviathan in the name of civil peace, the corollary of the biopolitical account of politics, power and war entails the acceptance of genocidal massacre in the name of life necessity:

> massacres have become vital ... If genocide is indeed the dream of modern powers ... it is because power is situated and exercised at the level of life, the species, the race, and the large-scale phenomena of population.
>
> (Foucault 1998: 137)

In sum, a population is not a flock in the ecclesiastical sense of the term, and it is not a people in modern political senses of the term.[12] Where the latter especially is concerned it is important to emphasise that a population is not comprised of subjects driven by different causations and expressing contrary wills. A population does not issue or obey commands. A population does not come to the bargaining table, and it does not become a political problem through inter-subjective disagreements. It does not pursue a manifesto and it does not issue political demands, negotiable or otherwise. The political regulation to which population may then be subject cannot therefore proceed through the pursuit and resolution of inter-subjective rivalries rendered tractable by the de-confessionalisation of politics and the privatisation of conscience, which historical manoeuvres are traditionally associated with the early liberal Civic Enlightenment of the late seventeenth century (Hunter 2006). Neither is a population a Westphalian (nation) state, geopolitically pursuing its interests in conflict and collaboration with other such entities.

It may well be the case that should a population find a voice – or be ventriloquised by political leaders – it will become a 'people', and in the form of a people pose just such inter-subjective problems. Foucault does not much go into this directly.[13] He does of course give very consistent attention to processes of subjectification and searches for alternative 'technologies of the self' in *The Hermeneutics of the Subject* lectures especially (Foucault 2005). In particular he was of course concerned with the complex ways in which liberal biopolitics, through the mechanism he calls governmentality or governance, posits autonomous domains of human behaviour governed by their own laws: such as those of 'economy', 'society' and more generically Life. He observes how in doing this liberal biopolitics also describes such autonomy as freedom and invokes that freedom as the principal mechanism by which individuals and collective groups are recruited into regulating and practising normalising self-governance on themselves (Foucault 2008).

In principle, then, a population is simply a cohort of individuals that arises in the context of some specific ordering of species existence. In relation to such ordering processes, it is not a matter of interrogating the will of the population or even of detailing the 'causal' laws to which it is subject. It is instead a matter of profiling the objective behaviour trafficking and transactions of cohorts of individuals, and how they are exposed to life events, through statistical analysis, probability analysis, market behaviour, distribution and risk management among other devices. When securitisation and risk became dominant regulatory practices as they did at the end of the twentieth century, population is commonly divided into risk pools whose risk properties become a new focus of scientific attention: risk embracing, risk aversion, objective risk subjective risk, social determination of risk and so on. That way, too, one man's debt becomes another man's income – and traded as such.[14]

As modern power learnt scientifically what it meant to be a living species in a living world – 'to have a body, conditions of existence, probabilities of life, an individual and collective welfare, forces that could be modified and a space in which they could be distributed in an optimal manner' (Foucault 1998: 143) – biological existence became the object also for political existence. Life entered the domain of power in a novel way because of the way in which life came to be understood in equally novel ways; not, *pace* Agamben, because it repeated a distinction drawn by Aristotle (Agamben 1998).[15] In addition to dealing with legal subjects, modern liberal regimes of power, therefore, began also to revolve around taking, 'life and its mechanisms', understood in these speciated ways, 'into the realm of explicit calculations and made knowledge/power an agent of the transformation of human life' (Foucault 1998: 143). Biopower thus first arose as that form of power which dealt with, 'the population, with the population as a political problem, as a problem which is at once scientific and political, as a biological problem and as a power's problem' (Foucault 2003: 245).

Race: from political arithmetic to demographics

> By Political Arithmetic, we mean the art of reasoning by figures, upon things relating to government … The foundation of this art is to be laid in some competent knowledge of the numbers of the people.
>
> (Charles D'Avenant, 1698)

Recall, again, that biopolitics is the expression which Foucault gave to a newly emergent set of power relations which he observed developing during the course of the eighteenth century. These began to revise the powers of police first instituted by the newly emerging monarchical states of the seventeenth century. Newer forms of political arithmetic and other power/knowledge devices began to change the ways in which species life was emerging as a referent object for modern power relations. Biopower was the label which Foucault applied to this changing, diverse and heterogeneous set of practices. It hardly requires restating, but in order that there should be no doubt about it, modern biopolitics is not something which is discovered in nature and it does not replace other forms of power. Neither, at least for Foucault, was biopolitics a base metaphysical distinction drawn by the Greeks at the beginning of modern philosophy, whose ramifications throughout the political tradition of the West reach a cataclysmic denouement, for Giorgio Agamben with Auschwitz. The manifold of modern power relations which attracted Foucault's analytic attention was not that. It was instead an historically specific manifold; one which was in fact also to become subject to continuous self-transformation and change because it was the expression of adaptive 'intelligent systems' rather than the expression of a metaphysically rooted juridico-political order.

The biopolitical mechanisms which attracted Foucault's attention were those revolving around progressive changes in the understanding of life as species existence. In particular, these included that early compiling of statistical data about the exigencies of life to be found for example in mortality tables such as those drawn up in England by Edmund Halley and John Graunt in the mid-seventeenth century. These began the process of providing what nowadays would be called policy relevant information about births, marriages and deaths, the size and growth of populations, the availability of utilisable man power and the incidence and distribution of disease; consequently, they were of direct interest to the newly emergent centralising monarchical states both in England and in continental Europe. But there was in fact nothing to prevent them being used to extend, support and develop commercial capitalism and the progressive social and political interests of its rising bourgeoisie as well – and they were.

For these practices to develop and become embedded as the means by which populations began to be regulated and governed required myriad changes in the very ways in which people, property, risk and contingency came to be regarded. Such technologies were not simply discovered. They were artifices engineered and re-engineered. In that complex historical process, the very weft and warp not only of government, politics, religion and commerce but also of life itself changed, as Life and the promotion of the living became the referent object of power upon whose higher resolution the development of *dispositifs de sécurité* of biopower relied. What it was to be a person, what it was to own a property in oneself according, for example, to John Locke, what it was to be contingent, what it was to capitalise, what it meant to be a biological individual, were all progressively redefined from the late seventeenth century onwards in ways that were to transform the societies, economies and politics of the entire Atlantic basin, and, through their imperial conquests, much of the globe.

In this modern interactive topography of biopolitical and geopolitical power relations, race was also a palimpsest of discursive practices.[16] Deriving from many different religious, political, economic and cultural encodings, that of police or *polizeiwissenschaft* was another.

A highly influential work of this kind, an exemplar of its type and especially instructive also for us here, in as much as it prefigures a lot of the biopolitical social Darwinism of the nineteenth and early twentieth centuries, was the six-volume *System for a Complete Medical Police* by Johann Peter Frank, a pioneering advocate of 'medical police' who held positions in Göttingen, Pavia, St Petersburg and Vienna during the course of the eighteenth century. *Medical Police* advocated a wide range of policies to improve the all-round health of populations including their physical well-being, strength, moral sensibility, civic discipline and 'beauty'.

Frank's work was vast in its range, incorporating prenatal and infant care, schooling, the regulation of housing, food and drink as well as the prevention of illness and the reduction of accidents and crime in its scope. Among its general objectives were the provision of a stronger tax base, increased yields in domestic productivity and a reduction in the number of destitute requiring public support (Eigen 2006). Albeit a product of the age of enlightened absolutism, and leaving aside the explicitness of its racial concerns, it is not unlike a contemporary party political manifesto. Indeed some of its explicitly racial concerns to do with improving the biological stock of the population are no longer as unfashionable now, in the age of a newly revived liberal eugenics, as they once were (Agar 2004).[17]

As Eigen notes, 'race' is, however, a large and capacious category. It is especially so in Frank's work. There, it is an equivalent term for a range of genealogically produced groups including families, inbred villages and classes of people, regional and ethnic groups and the species itself. Frank does not advocate 'a racism'. On the contrary: 'The idea that all so-called races are extended kinship groups – and that all kinship groups can be included under the loose category of race – is an unstated but necessary premise of Frank's work' (Eigen 2006: 187). Through it, he also promotes a proto-eugenic state-sponsored programme of restricted breeding practices to eliminate heritable diseases, mitigate individual suffering, to lessen the public burden of caring for the orphaned and infirm, 'and to proactively generate a healthier more productive population' (Eigen 2006: 187). There was, Frank insisted, no more important political task than this: the study of hereditary (today we would say genetics) and the translation of that knowledge into practical legislation. You may subtract the political rationality of race here, but what remains is a preoccupation with 'heredity' as great as our own, and a perfect expression of the racial logic of biopolitical governance in which politics becomes husbandry.

Advocating a form of human husbandry which could learn from animal husbandry – 'Would it not be wonderful if only a portion of the care devoted to the development of good animal breeds … would be applied to humans' (Frank in Eigen 2006: 188) – Frank's work powerfully illustrates where politics leads when it takes the life of species being for its referent object of power. For animal life is also species life. Preoccupation with the species existence of human existence quickly raises the issue of extra-human biopolitics in the form here, for example, of the inter-articulation of the human and the animal.[18] In the twenty-first century that extra-human biopolitics takes in every conceivable form of living assemblage – including viruses, both digital and molecular, and every form of life, not least those to do with mass food production.

Elevating illness, especially hereditable illness, to a category within human taxonomy, Frank was dedicated to the promotion of human welfare through its eradication via assisted breeding programmes, recommending that:

> beautiful people blessed with a strong, well formed and healthy body, even if they were robbed of all other resources, were supported in marriage with their equals in health and bodily perfection, and further supported in the raising of large families just like themselves, so that the number of strong, well-endowed body, citizens would eventually be increased.
>
> (Frank in Eigen 2006: 188)

In the process of developing and elaborating his schema for medical police, Frank also expressly recognised the need to subordinate religion, amongst other 'regressive' categories of social thought and practice, to the biopoliticised prioritisation of health and hereditary. In certain respects Frank may be said to have been 'socially progressive'. For example, he advocated healthy pairings without regard to social, economic or even racial category: 'Frank ventures further than most of his contemporaries in his thinking about the consequences and the benefits of racial mixing' (Eigen 2006: 189).

Individual races, in short, required the infusion of foreign blood if the overall vitality of the human race was to be promoted. Says Frank: 'The police must ensure that the human races are refreshed from time to time with foreign blood' (cited in Eigen 2006: 185). All this was, of course, nonetheless enframed in a general belief in European superiority. Again Frank is useful because he illustrates that even in the form of police a concern for population does not necessarily commit its practices to advocating any specific form of racism. Foucault argues that in the state racism of Nazi Germany it did result in such a position. But we miss the point of the very analytic of biopolitics, and in particular its relevance to the changing politics of race in our own times, if we fail to appreciate that biopolitics is not a specific racial doctrine. It does not require or promote a specific racism. It is not a doctrine or discourse of race. It is, instead, a complex of political rationalities and governmental technologies which require 'race' for their operationalisation.

Albeit an illustration of *polizeiwissenschaft*, what Frank's text helps to emphasise is the logic of human husbandry to which biopolitics became committed. One may, therefore, be misled, for example, if one confines oneself to asking about the cultural, religious and economic antecedents and drivers of specific forms of racism alone. One has to ask, in addition, how the specific mechanisms and techniques currently giving expression to biopolitical imperatives articulate race as a function of their everyday operational requirements. In the process of identifying optimal behaviour for a flourishing human species, biopolitically, specific individuals, types and groups are bound to be profiled as superior to, or nowadays more endangering than, others. Rank ordering of individuals and groups in terms of their fitness for life – 'social sorting' to use David Lyon's expression (Lyon 2003) – is as much a socio-technical as it is a political imperative of biopolitics.

The logic of the problematisation of governance and rule in terms of the promotion of life as species existence was, of course, also embedded within, and inter-articulated culturally with, other racially differentiating practices. Of special note, for example, was the way in which the term 'race' was adopted in cognate forms across Europe during the course of the eighteenth century. A complex set of philosophically articulated political rationalities ranged philosophically and theologically from Kant, Herder, Blumenbach and Meiners among others, to the epistemic commonplaces of medicine, geography, aesthetics, and philology. Race emerged here as a generically organising 'scientific' as well as political term (Eigen and Larrimore 2006; Appadurai 1996).

Race, therefore, is not a product of biopoliticised practise of rule alone. But the operationalisation of biopoliticised practices of rule does require race as one of its necessary mechanisms. If you biopoliticise you will racialise, however subtly or opaquely recoded in your biopoliticised operating technologies of power that racialising may be.

But, it was never in fact that subtle or opaque, most especially where it was used extensively to rationalise imperial conquest and enact colonial administration; for example, throughout the British Empire from the Americas to Africa and India (Stoler 1995; Sussman 2004; Houston 2003). Moreover the process operated as it were internally as much as it did externally. Initially, for example, it was critical to the consolidation of the restoration of monarchy in England and its allied conquest and plantation of Ireland during the course of the seventeenth century.[19] From the very beginning, in John Graunt's early compilation and analysis of mortality tables for London for example, 'questions of [political] order remain[ed] paramount' (Buck 1977: 72).[20]

Thereafter, during the course of the late eighteenth and nineteenth centuries its race character became even more evident as the European metropoles engaged in 'internal' as well as 'external' colonisation. Take this one example. During the nineteenth century, especially, theories of race were advanced both by the scientific community and by the popular daily and periodical presses in the United Kingdom especially. In much of the pseudo-scientific literature of the day the Irish in particular were regularly held to be inferior, an example of a lower evolutionary form, closer to the apes than their 'superiors', the Anglo-Saxons. Cartoons in *Punch* portrayed the Irish as having bestial, ape-like or demonic features and the Irishman (especially the political radical) was invariably given a long or prognathous jaw, the stigmata to the phrenologists of a lower evolutionary order, degeneracy or criminality.

'Thus John Beddoe, who later became the President of the Anthropological Institute (1889–1891)', records Anthony Wohl (2014)

> wrote in his *Races of Britain* (1862) that all men of genius were orthognathous (less prominent jaw bones) while the Irish and the Welsh were prognathous

> and that the Celt was closely related to Cromagnon man, who, in turn, was linked, according to Beddoe, to the 'Africanoid'. The position of the Celt in Beddoe's 'Index of Nigrescence' was very different from that of the Anglo-Saxon. These ideas were not confined to a lunatic fringe of the scientific community, for although they never won over the mainstream of British scientists they were disseminated broadly and it was even hinted that the Irish might be the elusive missing link. Certainly the 'ape-like' Celt became something of an malevolent cliché of Victorian racism.
>
> Thus Charles Kingsley could write 'I am haunted by the human chimpanzees I saw [in Ireland] ... I don't believe they are our fault ... But to see white chimpanzees is dreadful; if they were black, one would not feel it so much ...' (Charles Kingsley in a letter to his wife, quoted in L. P. Curtis, *Anglo-Saxons and Celts*, p. 84).
>
> (Wohl 2014)

Biopolitical power/knowledge has elicited enthusiasm and opposition in almost equal measure. For some it serves citizenship, for others it spells control (for example Maghiros *et al.* 2005). Here 'freedom of information' is confusingly and controversially combined with 'freedom through information' and 'freedom from information'. Surveillance via population census was, for example, resisted as much in the eighteenth century as surveillance by 'dataveillance' is resisted today. During the course of the violent debates over the 1753 Census Bill in England, for example, the charge was made that the compiling of census statistics by the Crown would result in reducing the people to the 'the numbered vassals of indiscriminating power' (Buck 1982: 33).

Conversely, however, statistics were not simply resisted. In the form of demographics they were also democratised and turned to liberally progressive causes promoting the interests of religious non-conformists and republicans as well as those of commerce and the rising bourgeoisie. A democratic interest superseded, or at least complemented, the continuing autocratic interest in the science of population. As statistics became democratised, from the mid-eighteenth century onwards, they became the preserve of non-conformist clergymen, physicians and those of a republican persuasion.[21]

Allied with such interests and causes, population science began to serve the cause of trade liberalisation and colonial expansion as much as it did social reform. Here political arithmetic became a demographics designed:

- to check the growth of the state;
- to demonstrate that according privileged political status to owning freehold in land was unwarranted;
- to demonstrate that mechanisms were available to secure more mobile forms of property (capital) which offered alternative claims to political significance and privilege; and
- to promote empire and colonisation in the cause first of civilisation and, later, of modernisation.[22]

In short, where once it was argued that property freehold assured independence from despotic state authority and encouraged republican virtue, political statistics documented changes in property – from land to capital for example – on the basis of which new political and social claims were advanced. By the end of the eighteenth century, armed with statistical procedures that promised to restore the propertied independence of all Englishmen, English political arithmeticians for example 'looked forward to counting citizens not subjects' (Buck 1982: 45).

Such counting nonetheless often also remained explicitly racial even where, for example, it was pitted against slavery and the slave trade (Appadurai 1996). It also led to the rank ordering of individuals and population groups in terms, for example, of what Benjamin Franklin in the eighteenth-century was to call 'improvability' and which late moderns call 'development' (Bowker and Star 1999). Sorting population groups, especially in relation to the improvability or developmental capacities, aptitudes and potentialities of concern to a progressively biopoliticised politics, necessarily also entailed demotion of some as biologically inferior to others. Thus eighteenth-, nineteenth- and twentieth-century biopolitics were racial in their own distinctive ways. These ways also inter-articulated with other racial instruments and doctrines at work as much in domestic as they were in imperial politics.

Demographics to biometrics: facial recognition and race

> Are we who we appear to be? All identity papers and, more seriously, every description of a person and all mug shots make it plain that recognition and identification are not the same.
>
> (Groebner 2007: 17)

Biopolitics are mobile and mutable. Populations change. So also do the means of enumerating, sorting and assessing them. Understandings of species life also change. So also do the means of intervening into and manipulating the life of living assemblages; most dramatically today through digital and molecular science. Bodies – individual and collective – are now conceived in terms of code, which code is manipulated through digitised information and communications technologies. As political arithmetic gave way to demographics, so also in response to these changes, and for the purpose of surveillance and regulation in particular, is demographics being supplemented by biometrics.

The events of 9/11, in particular, catapulted biometric systems to the forefront of the surveillance surge that accompanied the build-up of the war on terror. Such developments combine the politics of the body with the politics of population in a complex new topography of digitalised power relations. Governmentally, the body becomes a source of its own metric and populations are beginning to be aggregated and disaggregated in many novel molecularised and digitalised ways (Van der Ploeg 2003). This applies especially to the pervasive risk profiling of individuals and populations through socio-technical systems; systems which are as common in commerce as they are in the war on terror, promoting consumer capital, regulating population movement and controlling commerce and migration. Biometrics combined with the war on terror has turned making live and let die into a necropolitics of becoming-dangerous in which individuals and population groups alike are now scanned and surveilled, recorded and sorted for the purposes of assessing the risk which they pose across a wide variety of social, cultural, commercial, financial and terror domains. The very geography of danger has been transformed in the process. For reasons that I cannot detail here for lack of space, biopolitics of security in the twenty-first century has become governing through risk and the very cartography of danger has been transformed to include the virtual as much as the actual – the potential to become dangerous profiled by risk factors and not the actual commission of any crime. It gets lodged in the algorithms.

The biopolitics of governmental technologies is never, however, the politics of this or that technical artefact or system of auditing and account. Governmental technologies are socio-technical systems. Socio-technical systems are complex assemblage of codes, protocols, agreements, alliances, conventions, training programmes and so on (Law 1991; Law and Hassard 1999). The scopic regimes of their novel technical artefacts – for example the facial recognition systems, finger printing, iris scanning, gait recognition of biometrics – are deeply embedded in such assemblages. They do not simply work. Even when most automated, they are put to work – for example, through their computerised codes and algorithms:

> In these complex and dynamic socio-technical networks ATMs, doors, locks, keys, cameras, algorithms, etc. – function as political 'locations' where values and interests are negotiated and ultimately 'inscribed' into the very materiality of the things themselves – thereby rendering these values and interests more or less permanent.
>
> (Introna and Wood 2004: 180)

It is in their being put to work micro-politically that the neo-baroque politics contemporary biometrics do their sorting, and racial markers become embedded through the complex interface which such systems form between cultural encodings and technical systems.

Via their analysis of the most common biometric technique, facial recognition systems (FRS), Introna and Woods have provided a classic illustration of how this works. Detailing how FRS is one of a number of biometric systems, they examine how such systems operate through 'un-authored strategies' designed to regulate flows and access to flows rather than to manage and discipline human subjects directly. The operation of such systems is in a way 'un-authored' in as much as they are extensively automated with their scripts buried deep in their computer coding. Engaging with them, people are recruited into their pre-set and pre-emptive strategies for managing the conduct of conduct. Here sorting codes become silent, embedded within the socio-technical protocols of biometric systems that are difficult to avoid, challenge or revise. They do not simply empower, they enrol. People become enmeshed in their encoded strategisation of information, risk and management. Scrutinised by them, they in turn evade our counter-scrutiny. It is literally at the interface effected here between the social, the cultural and the technical by such systems that racial and social sorting now takes place biometrically.

With FRS certain other factors also apply. 'The targets do not need to surrender their face image, as they would their fingerprint, or their iris scan', for example (Introna and Wood 2004: 178). A face can be captured and (de)coded without the consent or participation from those being targeted. But this 'captured' image can only be useful if it is matched with an 'identifier'. Matching can be done by human beings or it can take place via algorithms. CCTV cameras may capture images but those images are scrutinised by security guards and police. Introna and Wood pursue the algorithm and explain the difference:

> If typists enter in the health details of a patient into the Health Service computer, then it is algorithmic to a limited extent in that software determines the extent of the information that can be entered, however it becomes what is usually understood as algorithmic surveillance when, for example, a program is installed which compares the patient records against signs of particular disease risk-factors, and defines or categorises patients automatically.
>
> (Introna and Wood 2004: 181)[23]

An algorithm is simply a mathematical, or logical, term for a set of instructions. Algorithms can be divided into trivial and non-trivial types, the former being sets of instructions that are only applicable to a specific situation, or a task that needs further explanation, the latter being instructions that will provide answers given any compatible input. Algorithms form the basis of modern mathematics and most importantly here, the foundation of computing. However, in themselves, algorithms are not accessible to computers, they need to be translated into a form that computers have been programmed to understand. This process, known as coding, produces software. Software is essentially composed of many coded algorithms linked together to produce a desired output from the hardware (Introna and Wood 2004: 180). However,

> software algorithms at the heart of facial recognition systems are propriety software objects. Thus, it is very difficult to get access to them for inspection and scrutiny. More specifically, even if you can go through the code line by line, it is impossible to inspect that code in operation, as it becomes implemented through multiple layers of translation for its execution.
>
> (Introna and Wood 2004: 183)

They are operationally obscure because of the ways in which they are embedded in technical systems; such algorithms are further obscured because, comprised also of very sophisticated statistical analysis, they are accessible only to certain categories of experts who are often surprised by the behaviour of their own algorithms (Phillips *et al.* 2003).

As Introna and Wood explain, locating a face image in the field of vision is not a trivial matter. For efficiency of operation and storage the face image is reduced to a numerical representation. In the process certain things must be overlooked or excluded. The image is sent to the software and the software determines if there is a match. If there is a match an 'alarm' is sounded to alert an operator to verify the match and initiate the appropriate action. The match could be a 'true' match or it could be a 'false positive'. A false positive means the algorithm made a mistake and the alarm would then be cancelled. There is a trade-off between false positive rates and recognition performance rates. If you are prepared to accept higher false positive rates then the recognition rate goes up. However, this gives you more false positives to deal with. Introna and Wood analysed the consequences of algorithmic exclusions and the ways the FRSs perform in practice.

They first did this through analysing the most extensive examination of FRSs to date, that of the Facial Recognition Vendor Tests (FRVT) of 2002 – independent tests sponsored by a variety of US federal agencies including the Defense Advanced Projects Agency (DARPA), the FBI and the State Department. The most surprising outcome for those involved in the FRVT tests was that the algorithms in use displayed particular identification biases. Recognition rates for males and old people, for example, were higher than for females and young people. They could not check for racial bias because the data set mostly consisted of Mexican non-immigrant visa applications. However, research by Givens et al. published in 2003 confirmed the biases established by the FRVT tests (except for the gender bias) and also found a significant race bias. The Givens analysis found that:

> Asians are easier [to recognise] than whites, African-Americans are easier than whites, other race members are easier than whites, old people are easier than young people, other skin people are easier to recognize than clear skin people.
>
> (Givens *et al.* in Introna and Wood 2004: 190)

These results were relatively small but when related also to how these socio-technical systems are operated culturally the results become magnified:

> it is not just the micro-politics of the artifact that is the issue. It is how these become multiplied and magnified as they become tied to other practices that is of significance ... We would claim it is here where the seemingly trivial biases may become very important as they become incorporated into actual practices.
>
> (Introna and Wood 2004: 191–2)

In short, when the alarm goes off the interface between the social and the technical is activated and existing biases against the usual suspects (including minorities) are triggered causing such groups to be subject to disproportionate amounts of scrutiny. Thus existing biases may be confirmed and even authorised by the system itself. Risk status is intensified by a combination of the operation and the operators of the system (Graham and Wood 2003). The operators may even override their own judgements as they may think that the system under such high conditions of operation must 'see something' that they do not (Introna and Wood 2004: 192). In the operation of socio-technical systems like biometrics, through which the biopolitics of the twenty-first century are becoming operationalised, you simply cannot remove the social from the technology or the technology from the social. It is the ensemble that matters.

Conclusion

Foucault's interest in security, race and war therefore arises out of, and remains integrally associated with, his changing analytic of modern political and power relations – the rules of truth and truths of rule which he calls politics of truth. It arises, especially, out of his concern with liberalism as a regime of power and in particular with liberal governmentality which takes species life as its referent object of power relations: that is, liberal biopolitics. In effect biopolitics makes war on behalf of its understanding of life. That war may be war in the traditionally understood sense of interstate conflict, such as the recent Iraq war, and here, especially, biopolitics may well be closely correlated also with geopolitics, since both forms of politics originated from the same crisis conditions which inaugurated the modern and have remained closely correlated with one another throughout the modern period. Foucault never had any problem understanding that the world is comprised of a manifold of changing kinds of power relations not least, but not only, those of modern *geo* and *bio* politics.

It is, however, plain from Foucault's lectures, in particular *Society Must Be Defended* and *Security, Territory, Population*, that the biopolitical war which liberal governmentality wages on behalf of life is conducted very largely also through the security discourses and practices through which it administers liberal peace. To repeat an expression used earlier, biopolitical peace is inscribed with the logic of war through the discourses and practices of security.

Precisely because biopolitics concerns itself with the regulation and promotion of life as species existence, and because the life of species existence is a changing manifold of power/knowledge, in which the technoscientific is wedded to the governmental, then species life must be assayed, sorted and classified if its existence is to be promoted. Some species existence is, for example, not amenable to the promotion of species existence. Not all species existence is promotable to the same degree. Some species existence may even be said to be antipathetic to the promotion of species existence. Biopolitics therefore requires mechanisms to differentiate species existence in ways that will usefully inform how it must best govern in order to pursue its vocation of promoting species life. Hence, while the project of making life live requires a letting die, often also an actively promoted killing, necropolitics, is the other face of biopolitics. This is precisely where race comes in for Foucault. Race is one of those mechanisms by which biopolitics adjudicates life for the purposes not only of saying how life is to be promoted but also which life has to be disempowered and disenfranchised in that cause. Race is a governmental technology as much as it is a biologised cultural and political ideology.

Biopolitics changes, however, according to what the complex assemblages of power/knowledge, which take species existence as their referent epistemic object, teach us about the changing conceptualisations and practices of species existence. As the political rationalities and governmental technologies of biopolitics changed so also did the ways in which race functioned as a biopolitical discriminator for the regulation of populations and the promotion of species existence: from political arithmetic through demographics to biometrics. However much it therefore appeals to the unity of the human race, biopolitics continuously finds itself in need of mechanisms which will, as Foucault puts it, break into the biological continuum in order to teach it how best to promote the race. While there is biopolitics there will always be race.

Notes

1 See Marks (2000). See also Elden (2006).
2 Classically in Sir William Petty's 1654 'Down Survey'. See the outstanding essays by Peter Buck (Buck 1977, 1982).
3 See, for example, in relation to US welfare policies, Cruikshank (1999).
4 See also Hinchliffe and Bingham (2008). On the post-human see Hayles (1999).
5 For a modern view see Sen (1999). For an account of the 'ethic' of transactional freedom see, for example, Barach and Elstrott (1988).
6 I make no apology for the length of the quotations.
7 All emphasis in these quotations here is added.
8 See in particular the Preface to Foucault (1989).
9 It is another matter to argue about the extent and degree to which modern politics – specifically modern liberal politics – is a form of biopolitics, indeed a form of global biopolitics. I am prepared to argue that but this was not Foucault's initial argument.
10 Foucault: '[t]his is the background that enables us to understand the importance assumed by sex as a political issue' (1998: 145).
11 My disagreement with Ann Laura Stoler's otherwise excellent, *Race and the Education of Desire* (1995).
12 It is not unusual for Foucault to observe that an older form of power relations continues to operate alongside a newer one. With very rare exceptions – notoriously, in relation to sovereign power – Foucault's analytics of modern power relations continuously stresses how diverse and heterogeneous they are, and how indebted they are to modern forms of knowledge. Foucault thus remained ambivalent about the fate of pastoral power: 'pastoral power … has never been truly abolished, and I am very likely still mistaken when I situate the end of the pastoral age in the eighteenth century, for in fact pastoral power in its typology, organization, and mode of functioning, pastoral power exercised as power, is doubtless something from which we have still not freed ourselves' (Foucault 2007: 202). My thanks to Julian Reid for drawing this quotation to my attention and for the intense discussion of pastoral and biopower which it stimulated.
13 Others do. See for example Rancière (1999). For a critical engagement of Rancière see Dillon (2005b). Slavoj Žižek acutely observes that Brecht was theoretically quite wrong to mock the East German *nomenklatura* facing the workers' revolt of 1953: 'Would it not be easier for the government to dissolve the people and elect another?' One should bravely admit, Žižek declares with evolutionary logic, 'that it effectively IS a duty – THE duty even – of a revolutionary party to "dissolve the people and elect another"' (Žižek 2006: 74).

14 I deliberately retain the term man here because it is a powerfully gendered business.

15 This is the point at which I can only register rather than detail my disagreement with Agamben.

16 On palimpsest and race see for example Rushdy (2001), especially ch. 1. On palimpsest and international relations see yet another pioneering essay by Michael J. Shapiro (Shapiro 2005). For an extensive account of the concept of the palimpsest, and in particular a wide-ranging theorisation of 'palimpsestuousness', in the context of contemporary critical theory and literary studies see Sarah Dillon (S. Dillon 2007).

17 James Mittra critically evaluates the UK House of Commons' Select Committee Report's case for a devolved, libertarian-inspired framework for the regulation of reproductive technologies (Mittra 2007).

18 There is in short an extra-human dimension to biopolitics which must deal with the interface of the human with other living assemblages. There is no time to go into it here but current preoccupations with the threat to humans posed by the transmission of bird flu, itself something procreated through the ways in which humans keep turkeys and chickens, is a good illustration. The human/chicken interface is a powerful stimulus to modern forms of policing. See Bingham and Hinchliffe (2008).

19 Originally and most notoriously through William Petty's infamous 'Down Survey' (Petty 1690). For a detailed account see Luis Lobo-Guerrero (2007: ch. 2). More generally, see Phiddian (1996); McCormick (2007). See also McLean (2004).

20 John Graunt's work was published as *Natural and Political Observations mentioned in a Following Index, and Made Upon the Bills of Mortality* (London, 1662). See the discussion in Buck (1977, 1982).

21 The passionate controversies surrounding the compiling of statistics in the eighteenth century prefigure debates about the growth of information and surveillance society in the late twentieth and early twenty-first centuries. On the one hand was the fear of the institution of a police state and on the other was the ambition to constitute a free and informed people capable by the very same means of contesting the powers of surveillance and centralisation and surveillance. For an introduction to surveillance studies today see, for example, Lyon (1994); Gandy (1993); Norris and Armstrong (1999); Caplan and Torpey (2001).

22 'Where one says civilization, the immediate implication is a will to civilize. This idea is linked to the passion to impose civilization on the Other' (Glissant 1997: 13).

23 Norris and Armstrong pioneered the term algorithmic surveillance in *The Maximum Surveillance Society* (Norris and Armstrong 1999).

Bibliography

Agamben, Giorgio (1998) *Homo Sacer: Sovereign Power and Bare Life* (Stanford: Stanford University Press).

Agar, Nicholas (2004) *Liberal Eugenics: In Defence of Human Enhancement* (Oxford: Blackwell).

Appadurai, Arjun (1996) 'Number in the Colonial Imagination', in Appadurai, *Modernity at Large: Cultural Dimensions of Globalization* (Minneapolis: University of Minnesota Press), pp. 114–36.

Barach, Jeffrey A. and Elstrott, John B. (1988) 'The Transactional Ethic: The Ethical Foundations of Free Enterprise Reconsidered', *Journal of Business Ethics*, 7(7): 545–51.

Bartelson, Jens (1995) *A Genealogy of Sovereignty* (Cambridge: Cambridge University Press).

Bingham, Nick and Hinchliffe, Steve (2008) 'Mapping the Multiplicities of Biosecurity', in Andrew Lakoff and Stephen J. Collier (eds), *Biosecurity Interventions: Global Health and Security in Question* (New York: Columbia University Press), pp. 173–93.

Bowker, Geoffrey C. and Star, Susan Leigh (1999) *Sorting Things Out: Classification and its Consequences* (Cambridge, MA: MIT Press).

Buck, Peter (1977) 'Seventeenth-Century Political Arithmetic: Civil Strife and Vital Statistics', *Isis*, 68: 67–84.

Buck, Peter (1982) 'People Who Counted: Political Arithmetic in the Eighteenth Century', *Isis*, 73: 28–45.

Caplan, Jane and Torpey, John (2001) *Documenting Individual Identity: The Development of State Practices in the Modern World* (Princeton, NJ: Princeton University Press).

Cruikshank, Barbara (1999) *The Will to Empower: Democratic Citizens and Other Subjects* (Ithaca, NY: Cornell University Press).

Dillon, Michael (2003) 'Virtual Security: A Life Science of (Dis)order', *Millennium – Journal of International Studies*, 32(3): 531–58.

Dillon, Michael G. (2005a) 'Cared to Death: The Biopoliticised Time of Your Life', *Foucault Studies*, 1(2): 37–46.

Dillon, Michael (2005b) 'A Passion for the (Im) possible Jacques Rancière, Equality, Pedagogy and the Messianic', *European Journal of Political Theory*, 4(4): 429–52.

Dillon, Michael (2007a) 'Governing Terror: The State of Emergency of Biopolitical Emergence', *International Political Sociology*, 1(1): 7–28.

Dillon, Michael (2007b) 'Governing through Contingency: The Security of Biopolitical Governance', *Political Geography*, 26(1): 41–7.

Dillon, Michael and Reid, Julian (2009) *The Liberal Way of War: Killing to Make Life* (Abingdon: Routledge).

Dillon, Sarah (2007) The Palimpsest: Literature, Criticism, Theory (London and New York: Continuum).

Eigen, Sara (2006) 'Policing the *Menschen = Racen*', in Sara Eigen and Mark Larrimore (eds), *The German Invention of Race* (Albany: State University of New York Press), pp. 185–202.

Eigen, Sara and Larrimore, Mark (eds) (2006) *The German Invention of Race* (Albany: State University of New York Press).

Elden, Stuart (2002) 'The War of Races and the Constitution of the State: Foucault's "Il faut défendre la société" and the Politics of Calculation', *boundary 2*, 29(1): 125–51.

Elden, Stuart (2006) 'Rethinking Governmentality', *Political Geography*, 26: 29–33.

Foucault, Michel (1975) *Surveiller et Punir: Naissance de la Prison* (Paris: Gallimard).

Foucault, Michel (1976) *Histoire de la sexualité*. 1: *La Volunté de Savoir* (Paris: Gallimard).

Foucault, Michel (1979) *Discipline and Punish: The Birth of the Prison* (Harmondsworth: Penguin).

Foucault, Michel (1986) 'Nietzsche, Genealogy, History', in Paul Rabinow (ed.), *The Foucault Reader* (Harmondsworth: Penguin), pp. 76–100.

Foucault, Michel (1989) *The Order of Things: An Archaeology of the Human Sciences* (London and New York: Routledge).

Foucault, Michel (1998) *The History of Sexuality*, vol. 1: *The Will to Knowledge* (London: Penguin).

Foucault, Michel (2003) *Society Must Be Defended: Lectures at the Collège de France, 1975–76* (London: Picador).

Foucault, Michel (2004a) *Abnormal: Lectures at the Collège de France, 1974–1975* (New York: Picador).

Foucault, Michel (2004b) *Naissance de la Biopolitique: Cours au Collège de France, 1978–1979* (Paris: Gallimard/Seuil).

Foucault, Michel (2005) *The Hermeneutics of the Subject: Lectures at the Collège de France, 1981–82* (Basingstoke: Palgrave Macmillan).

Foucault, Michel (2007) *Security, Territory, Population: Lectures at the Collège de France, 1977–78*, ed. Michel Senellart (Basingstoke: Palgrave Macmillan).

Foucault, Michel (2008) *The Birth of Biopolitics: Lectures at the Collège de France, 1978–1979*, ed. Michel Senellart, François Ewald and Alessandro Fontana (Basingstoke: Palgrave Macmillan).

Foucault, Michel (2014) *Wrong-Doing Truth-Telling: The Function of Avowal in Justice* (Chicago: University of Chicago Press).

Gandy, Oscar H. (1993) *The Panoptic Sort: A Political Economy of Personal Information* (Boulder, CO: Westview).

Glissant, Édouard (1997) *Poetics of Relation* (Ann Arbor: University of Michigan Press).

Graham, Stephen and Wood, David (2003) 'Digitizing Surveillance: Categorization, Space, Inequality', *Critical Social Policy*, 23(2): 227–48.

Groebner, Valentin (2007) *Who Are You? Identification, Dissimulation, and Surveillance in Early Modern Europe* (New York: Zone Books).

Hayles, Katherine (1999) *How We Became Posthuman: Virtual Bodies in Cybernetics, Literature, and Informatics* (Chicago, IL: University of Chicago Press).

Hinchliffe, Steve and Bingham, Nick (2008) 'Mapping the Multiplicities of Biosecurity', in Ali S. Harris, and Keil Roger (eds), *Networked Disease: Emerging Infections in the Global City* (Malden, MA and Oxford: Wiley-Blackwell), pp. 173–94.

Houston, Alan (2003) 'Population Politics: Benjamin Franklin and the Peopling of North America' (La Jolla, CA: Center for Comparative Immigration Studies, University of California, San Diego).

Hunter, Ian (2006) *Rival Enlightenments: Civil and Metaphysical Philosophy in Early Modern Europe* (Cambridge: Cambridge University Press).

Introna, Lucas D. and Wood, David (2004) 'Picturing Algorithmic Surveillance: The Politics of Facial Recognition Systems', *Surveillance & Society*, 2(2/3): 177–98.

Jacob, François (1989) *The Logic of Life: A History of Heredity; the Possible and the Actual* (Harmondsworth: Penguin).

Law, John (1991) *A Sociology of Monsters: Essays on Power, Technology, and Domination* (London and New York: Routledge).

Law, John and Hassard, John (1999) *Actor Network Theory and After* (Oxford and Malden, MA: Blackwell).

Le Goff, Jacque (1982) *Time, Work, & Culture in the Middle Ages* (Chicago, IL: University of Chicago Press).

Lobo-Guerrero, Luis (2007) 'Security, Insurance, and Power' (Lancaster University).

Lyon, David (1994) The Electronic Eye: The Rise of Surveillance Society (Cambridge: Polity Press).

Lyon, David (ed.) (2003) *Surveillance as Social Sorting: Privacy, Risk, and Digital Discrimination* (London and New York: Routledge).

McCormick, Ted (2007) 'Transmutation, Inclusion, and Exclusion: Political Arithmetic from Charles II to William III', *Journal of Historical Sociology*, 20(3): 259–78.

McLean, Stuart (2004) *The Event and its Terrors: Ireland, Famine, Modernity* (Stanford, CA: Stanford University Press).

Maghiros, Ioannis, Punie, Yves, Delaitre, Sabine, Lignos, Elsa, Rodriguez, Carlos, Ulbrich, Martin and Cabrera, Marcelino (2005) 'Biometrics at the Frontiers: Assessing the Impact on Society' (Brussels: European Commission Directorate-General for Energy and Transport).

Marks, John (2000) 'Foucault, Franks, Gauls', *Theory, Culture & Society*, 17: 127–47.

Mbembe, J.-A. (2001) *On the Postcolony* (Berkeley: University of California Press).

Mbembe, J.-A. (2003) 'Necropolitics', *Public Culture*, 15(1): 11–40.

Mittra, James (2007) 'Marginalising "Eugenic Anxiety" through a Rhetoric of "Liberal Choice": A Critique of the House of Commons Select Committee Report on Reproductive Technologies', *New Genetics and Society*, 26(2): 159–79.

Montag, Warren (2005) 'Necro-Economics: Adam Smith and Death in the Life of the Universal', *Radical Philosophy*, 134: 7–17.

Norris, Clive and Armstrong, Gary (1999) *The Maximum Surveillance Society: The Rise of CCTV* (Oxford and New York: Berg).

Petty, William (1690) *Political Arithmetick* (London: printed for Robert Clavel at the Peacock, and Hen. Mortlock at the Phoenix in St. Paul's Church-yard: London).

Phiddian, Robert (1996) 'Political Arithmetick: Accounting for Irony in Swift's *A Modest Proposal*', *Accounting, Auditing & Accountability Journal*, 9(5): 71–83.

Phillips, P. Jonathan, Grother, Patrick, Micheals, Ross, Blackburn, Duane M., Tabassi, Elham and Bone, Mike (2003) 'Face Recognition Vendor Test 2002', paper presented at the Analysis and Modeling of Faces and Gestures, AMFG 2003 IEEE International Workshop.

Rancière, Jacques (1999) *Disagreement: Politics and Philosophy* (Minneapolis: University of Minnesota Press).

Rushdy, Ashraf H. A. (2001) *Remembering Generations: Race and Family in Contemporary African American Fiction* (Chapel Hill: University of North Carolina Press).

Sen, Amartya (1999) *Development as Freedom* (New York: Knopf).

Shapiro, Michael J. (2005) 'The Demise of "International Relations": America's Western Palimpsest', *Geopolitics*, 10(2): 222–43.

Stoler, Ann Laura (1995) *Race and the Education of Desire: Foucault's History of Sexuality and the Colonial Order of Things* (Durham, NC: Duke University Press).

Sussman, Charlotte (2004) 'The Colonial Afterlife of Political Arithmetic: Swift, Demography, and Mobile Populations', *Cultural Critique*, 56: 96–126.

Van der Ploeg, Irma (2003) 'Biometrics and the Body as Information: Normative Issues of the Socio-Technical Coding of the Body', in David Lyon (ed.), *Surveillance as Social Sorting: Privacy, Risk, and Digital Discrimination* (London and New York: Routledge), pp. 57–74.

Wohl, Anthony S. (2014) 'Racism and Anti-Irish Prejudice in Victorian England'. The Victorian Web: Literature, History and Culture in Victorian England. <www.victorianweb.org/history/race/Racism.html>.

Žižek, Slavoj (2006) 'Afterword', to Jacques Rancière, *The Politics of Aesthetics: The Distribution of the Sensible* (London and New York: Continuum), pp. 69–79.

7

GOVERNING TERROR: THE STATE OF EMERGENCY OF BIOPOLITICAL EMERGENCE

Suam habet fortuna rationem
[Chance has its own reason]
(Petronius)

Introduction

In responding to contemporary terror there is an important political point to be made in first reminding ourselves that terror has long distinguished the ways in which the West has waged war and accounted for what it is to be modern. Neither is the peculiar terror of self-immolation in the pursuit of a higher cause confined to religious belief or peculiar to Islam. While the *passage à l'acte* of contemporary suicidal terrorism is a political phenomenon long nurtured in the many protracted local and global struggles through which Islam and political modernity have been encountering one another, terror has long figured also in Western philosophy as a generative principle of formation for modern political self-hood. There is, therefore an important philosophical point to be made in reminding ourselves that the question of terror has never been far from the very formulation and pursuit of the question of Being in western thought as well as the question of politics in Western practice (Strauss 1998). Hegel's analysis of the dialectic of secularisation placed terror at the very heart of the modern political experiment, for example, and Heine quipped that Kant had far surpassed Robespierre in intellectual terrorism. I shall return to the Kantian terror which motivates liberal internationalism towards the end of this chapter.

Every 'terror' thus has both a philosophical as well as a political register, finding its expression on the one hand as a political practice and on the other as a philosophical problematic. Taking the very positivity of contemporary terror as its provocation, this chapter does not provide a culturally specific genealogy, political description or rigorous philosophical audit of the terror of contemporary Islam or that of modern Western philosophy and politics. It recognises the importance of these tasks and the focus it takes is not meant to diminish the need for either of them. Its purpose is, however, different. That purpose is signalled by the deliberate ambiguity registered in the title of my chapter.

Governing terror

Governing terror first references the massive global security effort that is now devoted to the war on terror. Its primary purpose is to bring terror within the political rationalities and calculative control of Western security technologies with the aim of destroying it; or reducing it to 'manageable' proportions. In that sense the aim is to eliminate terror through the advance of good government or make terror at least governable through the advance of security technologies. In the process there has been a massive extension and intensification of the political rationalities and governing technologies of security into almost every aspect of western life. At the very same time, however, *governing* terror also signals the degree to which Western policies, local and global, are themselves also determined by a widespread fear of terror. This radical ambiguity – Western societies themselves governed by terror in the process of trying to bring terror within the orbit of the political rationalities and governmental technologies now dominant in the West – calls for a continuous double reading of terror. It also betrays a profound suspicion that the more effort that is put into governing terror the more terror comes to govern the governors and their governing technologies.

In certain respects, this is not a new thesis. Geopolitical analysts regularly note the danger of being dominated practically as well as psychologically by the strategies of one's enemies. But I wish to state the thesis quite differently because I think the reasons for it are as much biopolitical as they are geopolitical. We do not fully appreciate the extent to which liberal societies are themselves governed and seek to govern globally through what, in the process of interrogating the mechanisms by which the basic biological features of the human species became the object of its political strategisation, Michel Foucault called biopower. We do not widely understand the kinds of imperatives to which the biopower of biopolitics now orders the political rationalities and governmental technologies of the West. Neither is it widely appreciated to what extent, and how, Western biopolitics simply is a diverse and changing collection of *dispositifs de sécurité*; *dispositifs* which also came to revolve around a kind of low-intensity but all-pervasive terror of contingency long before the contingency of global terror entered the scene. Thus it is that the contingency of global terror resonates powerfully with the terror of global contingency to engender a dangerous hyperbolicisation of security and fear in the West which widely amplifies as it circulates and responds to that posed by the threat of Islamist terror.

By 'security' Foucault did not mean a universal value, or condition of possibility, for a political subject. He meant a certain set of mechanisms through which species life is regulated. Moreover that set of mechanisms is itself governed by certain key analytical categories foremost among which is contingency. For the moment the statement of my thesis is simply put. *The war against terror emerged out of a generic biopolitics of contingency in the West, and is being conducted according to its political technologies and governmental rationalities, as much as it was precipitated by a contingent terroristic event directed against the epicentre of geopolitical hegemony in the United States.* The biopolitical processes involved have been underway for some very considerable time. They were not initiated by the attacks on New York and Washington in September 2001. These attacks were driven by a complex of geopolitical factors both local and global. But those attacks also amplified and intensified a generative principle of formation which has long governed the biopoliticised security technologies of the West. That generative principle of formation is contingency itself. And it is here that the radical ambiguity of my title emerges most powerfully. What most characterises global terror we are persistently told is the very certainty of its radical uncertainty. We don't know when terrorists may strike, we do not know how they will strike, and we do not know with what terrifying effect they will strike. We only know for sure that they will strike. It is the very contingency of terror that distinguishes its operational practice. But contingency is similarly also the very operational heart of the security *dispositifs* of the biopolitics of security as well.

It is neither geopolitics nor biopolitics, alone, but the toxic combination of the two which now drives Western security practices. What contemporary terror serves to amplify and intensify is the radical contingency on the basis of which biopoliticised life globally has long been understood to operate in Western security politics. These now translate risk and chance from the celebrated occasioning of the dynamic socio-economic, technoscientific and political supremacy of the West into the occasioning of terrifyingly dangerous uncertainties amplified and circulated by its very own forms of existence. The contingent which now governs Western life thus radically subverts what it first made possible – liberal biopolitics as such – widely circulating and intensifying its security neuroses.

How biopolitics operates is also simply put, and it is as well to put it now. Biopolitics is a collection of *dispositifs de sécurité* which secures – that is to say regulates, strategises and seeks to manipulate the circulation of species life – by instantiating a general economy of the contingent throughout all the processes of reproductive circulation which impinge upon species existence as such (Dillon 2007). How it does so, complementing the contingency of global terror with the terror of generic contingency, is the puzzle this chapter seeks to explore. It is a puzzle that extends far beyond any sociological account of risk society. From the perspective of this chapter the commodification of the contingent as risk, for example, is not an epiphenomenon of the social giving rise to something called 'risk society'. The contingent has been the very principle of formation for the social and the political alike for some considerable historical time, giving rise to an all-pervasive biopolitics of security in the West whose ontopolitical condition of operationality is that of contingency itself. I emphasise that this biopolitics of security revolves around the governance of the contingency which attends all the dangers that arise from the very ways in which the circuits of species life locally and globally function with increasing pace, unpredictability and density of connectivity.

The contingency around which biopolitics revolve has been evolving since the beginning of the modern age. It includes within its compass almost every artifice of Western life, from capital accumulation, financial flows, information and communication systems, to business continuity, healthcare, container shipping and port management, from food chains and energy grids to counter-terrorism, globalised criminality – especially in people, drugs, sex, organs and many other illegal substances – and pop-up warfare, from trans-continental tourism, the design of street architecture, risk-based governance of life assurance and pension funds to school outings and nursery provisions. The contingent – perhaps the trope of our neo-baroque politics – now rivals the economic and the social. It has achieved the status of an independent field of formation to whose political rationalities, governmental technologies and operational devices the social and the economic must now also submit. Contingency's very trade is that of insecurity. Hence, if the government of terror has become one of the single most important principles of formation around which Western security politics and the transformation of Western societies now turns, the terror of contingency predates it in a geopolitically warranted global biopolitics of security that has long come to wager the life of the species on its own governmental technologies and political rationalities.

Dispositifs of security

Recall that different problematisations of security are comprised of different discourses of danger. Different discourses of danger revolve around different referent objects of security, such that different referent objects of security give rise to different kinds of governmental technologies and political rationalities. 'Security' is therefore inscribed as a problematic before it gets inscribed as a value, a policy or a politics. The problematic of security posed for example by Life, is simply not going to be the same as that posed by territory (*la patrie*), sovereignty (*volk*, *reich*, *führer*, *demos*) or, indeed, 'reason' (*logos*, *raison*, *rationalität*). Employing life as the referent object of security and governance, security managers (a revealing term) are beginning these days to write policies and trade securities much like underwriters and stock analysts. Actuaries and traders in futures, they deal in the calculus of contingency, risk, uncertainty and probability at least as much as they do in the geo-strategic calculus of state policy and sovereign wills (Dillon 2006).

Historically, modernity has been distinguished by at least two great *dispositifs* for the problematisations of security. The one, revolving around the referent object of sovereign territoriality, has been the geopolitics of security. The other, revolving around the problematic of Life, specifically addressed in terms of population, has been the biopolitics of security. These arose early in the emergence of the modern age. They have coexisted throughout the modern age. And they continue to mutate in correlation with one another in late modernity. Foucault began their story (Foucault 2003, 2004b, 2004c). But Western biopolitics of security entered a new chapter in the twentieth century. It did so because technoscientific advances in the life sciences combined with massive changes in the demographics of Western populations, such as those concerning specifically, for example, their health and morbidity, transformed the politics of life in the West as much they transformed Western understanding of what it is to be a living thing. One can therefore say quite simply that the life which biopower takes as its referent object of power, a life which Foucault first interrogated in the form of population, has been transformed by the history of the twentieth century. In the process of transforming biopolitics, the history of the twentieth century has also transformed its intersection with geopolitics. What cannot be equally simply said, however, is how the West now understands what it is to be a living thing, how that transformation has impacted on the political rationalities and governing technologies of biopolitics and how the very correlation of geo- and biopolitics has mutated in the process. Only certain aspects of that complex of problematics can be explored here.

It is important to recall briefly how geopolitics of security differ from biopolitics of security. They differ in terms particularly of their referent objects of security, their political rationalities, their security mechanisms or technologies, their different histories and accounts of power relations, and, finally, their different ontological and epistemic assumptions. In Foucault's terms these *dispositifs* of security are quite different discursive ensembles. They talk a different security talk; they make visible in different ways; they are comprised of different artifices of government and rule as much as politically inspired killing. And so, they now also specify, of course, novel rules of truth as well as truths of rule.

On the one hand, there is sovereign juridical power over death. On the other, there is biopolitical power over life. On the one hand there is territoriality and the geopolitical mechanisms of war, diplomacy alliances, subjective self-interest, *raison d'état* and *macht politik*. On the other, there is power/knowledge, governmental technologies and political rationalities. On the one hand the history of the present in terms of capitals, chancelleries, empires, states, nations and ideologies, including also their spectacles of power and war. On the other, the history of the present in terms of the changing power/knowledges of governmental technologies and institutions beyond traditional politics and the state including, for example, medicine, psychiatry and confinement. This governmental power reaches far and wide into the everyday ordering of diverse individual subjectivities, populations and 'things', ranging from the corporeal sexual and psychic now into the molecular and the somatic (Rose 1999, 2001, 2003; Novas and Rose, 2000), scaling extensively down also through the micro and on into the nano (McNaghten, Grove-White and Wynne, 2004). On the one hand pre-formed subjects of will exercising power they are said to possess in calculative pursuit of their interest – above all, and foundationally, that of security. On the other, subjectivities whose identities, constitution and empowerment are conjured-up or secured by the very power/knowledge relations, governmental technologies and political rationalities that the subjectivist orthodoxies of geopolitics blithely tell us they freely exercise.

As a changing collection of *dispositifs de sécurité*, biopolitics not only functions through mechanisms that elevate contingency, in particular, into a dominant field of formation for Western societies as a whole, it similarly also opens up an entirely different spatial, scopic and theatrical configurations of security. If 'distribution' is the spatial figuration that characterises traditional geopolitical rationalities and technologies of security, 'circulation' is the spatial configuration which characterises the biopolitics of security. Whereas 'distribution' signals a world understood to be divided between sovereign territorial political subjects and their competing hegemonies, 'circulation' concerns a world understood in term of the biological structures and functions of species existence together with the relations that obtain between species life and all of its contingent local and global correlations. If geopolitically driven imperialism seeks to control the distribution of territory and resources, biopolitically driven imperialism seeks to control circulation as such. While most security analysis focuses upon the geopolitics that has committed the West – notably the US and the UK – to self-defeating strategies of global intervention and pre-emption,[1] this chapter interrogates the biopolitical imperatives at work in the closely associated hyperbolicisation of security which is so profoundly subverting the democratic politics and institutions of the West.

The single most important difference, the one under which all these others assemble is, however, the shift in the referent object of security from sovereign territoriality to Life. For once you make Life the principle of formation around which the problematisation of security, fear and danger revolves then the politics of security are transformed. Most notably they become subject to the changing ways in which the life sciences – in particular biology – specify what life is. From the outset, then, we have to draw a simple, almost banal, distinction between discourses of the human and the discourses of life. In our tradition, beginning classically with Aristotle, the discourse of life has never been confined to the human (Lennox 2001). Conversely, beginning again also with the Greeks, the discourse of the human has never been confined to mere life. However problematically life and the human may be related, and their relation is as problematic as the terms themselves, they are not the same thing. Indeed the very referent object of contemporary biopolitics of security, Life, is undergoing an unprecedented transformation in response especially to the confluence of the molecular and digital revolutions. Here, under a rapidly evolving bio-economical regime, we are dealing with a biopolitical imaginary of Life which understands life to be an adaptive and emergent process of non-linear adaptation and change.

Threat perceptions organised according to the security codes of geopolitics similarly also differ from those induced by a concern for Life and the promotion of species existence. The logic of threat installed by liberal biopolitics of security is ultimately not that of an externalised enemy alone. Neither another competitive state, nor an existential other form of Life, the threat to Life in the liberal struggle biopolitically to secure Life becomes life itself – the very means by which life-like properties circulate and propagate. The threat to Life that life itself poses is also an infinitely adaptive and changing one because adaptive emergence has itself become the very vital sign of life. One no longer asks if something is alive. One asks, is information exchange still happening here, and is a body capable of moving out of phase with itself through a combination of its own recombinatory genetic processes and correlative stimuli from its environment to produce further morphogenesis? Life-like properties characterise digital as well as molecular viruses, for example, as much as they do human beings. In the process, the project of 'making live', as Foucault put it, is compelled to take on novel functions of correction, punishment and death and deploy them violently against life on behalf of Life. For a life that does not display adaptive recombinatory morphogenesis may be non-life threatening life itself.

You cannot secure anything unless you know what it is. Integral to the problematisations of security are the ways in which people, territory and things are transformed into epistemic objects (Rheinberger 1997; Knorr-Cetina 1999). An equally profound shift in the field of power/knowledge therefore follows a shift in the referent object of security. What arises is not simply an epistemological adjustment that liberal biopolitics brings to the geopolitics of security; as if the one was only making up for a contingent lack in the other, supplying additional knowledge of yet another feature of security. An entirely different field of formation for security knowledge arises in the biopolitics of security; not some adjustment of the cognising political subject of geopolitical power, but a shift both in the very object of cognition – to what it is to be a living thing – and the epistemic practices of political cognising, focusing on the heterogeneous diverse and unpredictable dynamics that characterise the circulation of things that display life-like properties.

The biopolitical emergency of emergence: a new 'real'

> Between the already encoded eye and reflexive knowledge there is a middle region which liberates order itself.
>
> (Foucault 1989: xxii)

Life (like Being) is said in many ways (Aristotle 1986a, 1986b). When Foucault first talked about the advent of biopolitics in the modern age, the biopolitical referent for life which attracted his attention at that time was 'population'. As noted previously, population is not a people. People are formed by some combination of religious, racial, linguistic, cultural, social, economic or political ligatures of belonging. Population serves, instead, as a datum. As a datum it was initially the referent object supplied in particular by statistical analysis of productive and reproductive properties belonging to a cohort of individuals as well as the incidences of risk and danger to which they might be subject. Population, Foucault carefully emphasised, was in essence an aleatory phenomenon. Contingent upon the principle of formation that happened to pull a cohort of individuals together, population also displayed many different aleatory features, for example, initially, in the early political arithmetic of statistics, those of births, marriages and deaths (Hacking 1982, 1990).

The changing reality of life

This chapter takes its particular inspiration, however, from Michel Foucault's pithy observation made in the Preface to *The Order of Things*, in which he says that, 'between the already encoded eye and reflexive knowledge there is a middle region which liberates order itself' (Foucault 1989: xxii). It is that middle region which preoccupies me: neither the cultural specificities or political character of the current deployment of suicidal terror attacks, nor the terrorising shock and awe of Being that has concerned modern reflexive thought. It is the 'middle region' which Foucault posits, where our changing discourses of fear, danger and threat are engendering a new principle of formation and new modes of operability derived from an account of the 'real' founded in changing understandings of species life that is beginning to transform the cultural and political codes of security – civil and military, a distinction long difficult to maintain – as much as they do the epistemological and ontological assumptions upon which problematisations of security and war, in terms of sovereign subjectivity, were traditionally once based. In transforming our security codes, this newly emergent social and economic 'real' founded in a transformation of species life and all the circulatory mechanisms which characterise it, is transforming the regulation of life locally and globally.

It turns out that life-like properties exist everywhere these days. These are regularly now installed not only in weapon and surveillance systems but also in what were once thought to be the most inanimate of substances such as walls, metals and plastics. It is precisely here in the ground of life itself that contemporary biopolitics of security therefore intuit 'a pure experience of order and of its mode of being' (Foucault 1989: xxiii) radically different from the Newtonian physics of a mechanistic and positivistic 'real' that once inspired the West's traditional state-centric territorial geopolitics of sovereign subjectivity. For the burgeoning molecular sciences of life, and their allied digitalised sciences of animation, every thing is capable of being connected to every thing else informationally. This seems to have been a lesson first taught by biosemiotics and cybernetics (Sebeok 1999). Construe animate and inanimate material in terms of information and you connect them up through feedback loops of informational exchange, whose positive non-linear outcomes became vastly more interesting and important than the supposedly fixed properties of living systems, because they did not simply pose more interesting problems, they also offered the prospect of more creative self-organising problem-solving and adaptive behaviour (Mackay 1969; Kay 1993, 2000; Hayles 1999; Doyle 1997, 2003; Thacker 2003; Sebeok 2001). Here in liberalism's digitalised and molecularised, as well as globalised, biopolitics of security, the problematic of knowing is experienced differently because the problematic of life appears to impose itself differently. And, when the problematic of life imposes itself differently in terms of a changing order of the real, the problematisation of politics and security follows suit.

The intense subject of every conceivable kind of technoscientific investigation, the biological life of the life sciences and the animated life of digitalisation is not, however, a subject. Whatever else it may be, it is commonly now agreed by life sciences, both 'natural' and 'artificial' (a problematic distinction), that life is a process. Securing processes that constitute mutable and adaptive 'bodies' of every description – military strategic discourse of the Revolution in Military Affairs, for example, refer to such mutable military bodies as swarms (Dillon 2002b; Arquilla and Ronfeldt 2000, 2003; Edwards 2000) – poses different challenges and calls upon different power/knowledge formations than did those devoted to securing subjects assumed to be bodies enacting their wills.

In this emergent middle region that is changing the empirical orders prescribed for it by its primary security codes, liberal Western security politics – its changing institutions, military strategic discourses, technoscientific establishments, security mechanisms and practices – is induced by its preoccupation with life simultaneously also, therefore, to appeal to a new ground, a new philosophical basis, for the problematisation of security, fear and danger in life itself. Indeed in the very capacity to induce, manufacture, engender, distribute and disseminate as well as exploit and deploy life-like properties in and through all manner of material in its martial rage to enhance Life at the risk of life itself. These themselves depend upon a radicalisation of the contingent down into the molecular structure of morphogenesis itself, and out into the wider capillaries of existence that comprise the complex networks of global life. Consider just one example typical of many provided by Sir David Omand, at that time Intelligence and Security Coordinator to the British Cabinet Office preoccupied with the reorganisation of civil contingencies and national resilience in the UK following the 9/11 attacks:

> There are certain obvious characteristics we need to take into account in our planning. The speed and penetration of global communications. The tightly coupled markets that can transmit shocks instantly around the globe. The known vulnerabilities of complex information infrastructure, for instance controlling logistic systems or power grids. More fundamentally the commercially competitive pressures on the Boardrooms that now control most of our critical national infrastructure that in years gone by would have been in public sector control or at the least subject to influence in the public interest. Just in Time value chains, leanness and speed to market all can introduce greater fragility in the face of unexpected disruption. Our knowledge of these inter-relationships is far from complete. I know of no full mapping of an advanced economy anywhere in the world, or even of a manageable methodology for obtaining one.
>
> (Omand 2005).

Biopolitically speaking, Life has thus become defined in terms of an 'open system' engaged in transformative, informationally driven and knowledge-sensitive exchange with its 'environment'. An intelligent system not only capable of adaptive learning but defined (and measured) as a living thing, the vital signs of life become those of the informationally defined capacity to adapt. Living things so described as contingently performative as well as situated, and co-evolving with their environments, are not only said adaptively to govern themselves. They are emergent. Emergent means that they are capable of moving out of phase with themselves and becoming other than what they were. Beyond the life category of population, albeit population science continues to be a very important aspect of the life sciences, informationally driven self-organising self-propagation is now the object of life science, because that is what the life sciences now teach us life consists in. Or to put the matter differently, population is now informationally identified; a function of codes, even of biometric bar codes (Introna 2005; Introna and Wood 2004).

An important but subtle shift from Foucault's initial account of the self-governing freedom characteristic of disciplinary anatomo-politics and regulatory biopolitics takes place here. It has profound impact in elevating the very status and register of the aleatory in biopolitics, the biopolitics of security in particular. That shift occurs in this way.

With open systems, agency becomes distributed. No longer centred in the subject, cardinal amongst whose fixed properties was said to be freedom of the will and a property in its body, or indeed of the subjectification of the subject via the self-administered disciplinary norm – albeit these continue to operate ubiquitously as governmental devices – open system agency is a function of the disseminatory properties of the network character of the open system itself; that complex adaptive assemblage of informational interchanges and transactions between system and environment, that define the very openness of the system. If there is a norm here, the norm is not a statistical or disciplinary one in the sense that Foucault first proposed in his panoptical account of power/knowledge relations. It is the performative norm of emergence, adaptation and transformation whose criterion of fitness is signalled by the term 'resilience'; now used widely in the UK's response to this revised biopolitical re-problematisation of security with the institutionalisation of a Civil Contingencies Secretariat in the Cabinet Office, in July 2001, tasked with coordinating the UK's response not simply to Islamist 'terror', but any and every perturbation that may threaten the systemic continuity of the complex of local and global networks and infrastructures that situate and sustain life in the UK. Biopolitics of security have not only gone global they have also gone molecular and recombinant.

The changing reality of chance

Contingency, too, is said in many ways. From the Greeks through the Romans to the modern political imagination first inaugurated by Machiavelli, contingency has had a history and thus 'chance', according to the Latin tag from Petronius' *Satyricon*, 'has its own reason'. If contingency has always had its reason, biopolitically that reason has become reliant also upon the immanence of a pure operationality constantly strategising without significance (to adapt Agamben's account of modern law). To resist that liberal life reasoning is to join an evil insurgency against the principle of both reason and life which liberal biopolitics of security seek in their turn to propagate with the viral moral force of full spectrum dominance. Its quasi-transcendentals of circulation, connectivity and complexity are simultaneously both warranted and subverted by the very figure in which they find their cohesion and coherence. What the contingent makes possible, it also makes impossible. Thus the pure operationality demanded by the radicality of contingency that nonetheless results in the emergency of emergence becoming the norm for a biopoliticised life striving for that ever receding fitness for purpose of adaptation itself which complexity theorists claim to have found at 'the edge of chaos' (Gleick 1987; Kellert 1993; Gell-Mann 1994).

Chance thus always derives its meaning from the wider system of cultural signification – including also the very understanding of signification as such – within which it arises. Translated from the realm of fate or Fortuna, chance entered the signifying domain of mathematics with Pascal and the discovery of probability (Daston 1988; Kruger *et al.* 1987; Hacking 1995). Thereafter, with the growth of statistics and the development of statistical science, chance became risk (Bernstein 1998; Taleb 2001). It was not simply the chance to which somebody, something or some course of events was exposed affecting its practices as well as its price, risk itself was eventually also to become a product that could be bought and sold. Risk commodified chance, making it capable not only of being calculated but also of being traded. In that sense risk does not conform to Beck's account of it as an epiphenomenon of a certain social form: that of modern post-industrial society. Beck argues that the social form cannot sustain the risks that it has engendered. Almost all of his examples, however, prove the contrary. Insurance has adapted to and seems capable of governing the very unsustainable risks – terror amongst them – that Beck cites (Beck 1992; Erickson and Doyle 2004). But my argument is not especially with Beck's risk thesis. Contingency for biopolitics is not the epiphenomenon of a social form. Liberal governance, articulated biopolitically, grounds its freedoms in an ontological contingency of species being newly understood and newly experienced socially, politically and scientifically as radically contingent because ontologically emergent.

For with life understood as the contingency of distributed agency, it is neither the capillaries that transmit and disseminate the informational exchanges of which life-like properties are now said to be comprised, nor the nodes that connect up the capillaries, which determine the essential attribute of networks of life-like things; though a claim is regularly entered for both features. Ultimately, it is the capacity to respond to the moment of contingent interaction. It is the event itself that defines and measures their openness. It also closes them to considerations other than performative excellence in the management of the event; a feature commonly described as fitness or resilience. Thus has the good been comprehensively technologised as contingent operational tacticity.

What matters most in this ontologically 'revitalised' and epistemically revised, biopolitical account of 'the real' is the moment of becoming itself; to the degree that it also reflects a Spinozist or Deleuzean enframing one might quip, 'becoming-itself' (de Landa 2002; Protevi 2001; Bonta and Protevi 2004). That moment is a conjuncture – Althussereans might say an encounter (Althusser 2006) – the product of an intersection or complex of connections. In short, connectivity effects criticality, the moment of emergent becoming. No connection no becoming. No becoming-connected, no life [sad bastard, or failed state]. Life understood within this informationally expressed ontology of Being as Becoming not only elevates 'connectivity' into a principle of emergent life formation, but 'circulation' is similarly endowed with such a new and radically extended register of significance as well. For what circulation does is effect connectivity. Connectivity is a function of circulation. Conjunctures – of events, transformations and phase changes – are effected by a radical connectivity of circulating forces, elements and circumstances. These combine to effect novel outcomes whose very 'eventuality' is not itself subject to linear prediction.

The conjunctures of circulating connections are, in the new language of networks, 'complex'. Things that are complex are more than the sum of their parts. Thus emergence is both example and function of complexity. For what emerges is not, however complicated, an arithmetic product of the sum of parts – of elements connected and combined because their circulation has brought them together at a particular moment, in a particular way, at a particular place. The very singularity of the conjuncture affects an outcome that could not have been anticipated for it, or the sum of its parts. In fact such a thing meets the age-old definition of crisis as moment of danger and of opportunity. It contemporary terms, despite the weight more often placed on it as an occasion of danger rather than of profit, that moment is the moment of risk. In short, connectivity and circulation produce complexity. Complexity is a product of a connectivity that circulation effects whose initial conditions cannot be established with certainty. Alternatively we can say that complexity is what happens when connectivity and circulation combine to effect singular encounters whose outcomes display properties incapable of being predicted in advance of the moment of conjuncture itself, because their origins remain, however minutely, opaque (Dillon 2005). In complexity speak this is known as 'sensitive dependence on initial conditions'. Additionally, encounter also entails invention and this compounds the unpredictability – others would more accurately say, undecidability – to which such systems are said to be subject. In politics such conjunctures might be labelled crises, revolutions or catastrophes. Dealing with them increasingly defines modern politics. Indeed, since its advent with Machiavelli, it defines the evental temporality of political modernity itself (Vatter 2000; Althusser 1999).

Some acute philosophers and cultural analysts noted this almost a century ago and observed how the state of exception had become the politically modern norm (Benjamin 1996). My argument contrasts that state of emergency born of a juridico-political analysis of sovereign subjectivities, however, with the norm of a state of emergency born of a contemporary biopolitical analysis of emergent life. Thus the contingent concatenation of forces operating in the moment – bearing on the emergent evental character of living things continuously in-formation – is complemented by the corresponding importance now attached to the very disseminatory processes by means of which the conjuncture of living events, of 'evental' life, is continuously effected. In short, living things are said to be alive here to the extent that they continue to be distinguished by emergence. Emergence has become *the* vital sign of life for the global liberal biopolitics of the digital and molecular age. Key to having life-like properties is this capacity for emergence. The event of living things biopolitically is evental. Ontologically speaking, connectivity, circulation and complexity begin to acquire the status of quasi-transcendentals for life understood as evental. Epistemically speaking, the creative correlations of circulation, connectivity and complexity begin to compose a new science of life. In the process the figure of Man remorselessly recedes as this new life science progressively overwhelms the human sciences of which it was once said that the figure of Man was comprised.

What lends coherence, however, to these emergent quasi-transcendental principles of circulation, complexity and connectivity, which characterise the newly 'vitalised' biopolitics of the twenty-first century, and in particular lends coherence to their epistemic 'objectness', what draws them together so that they begin to constitute a newly articulated plane of formation comprised of its own autonomous epistemic laws and dynamics, is a new 'reality' possessed of its own 'objectness' in relation to which the social, the economic, the cultural and the political real are progressively being revised. That new real is in many respects the old real, the contingent itself, but it is an old real newly articulated in molecular and digital terms that now forms a widespread popular discourse of management and policy-making throughout all aspects of Western life.

The emergency of biopolitical emergence

Beyond mere risk – the product of an earlier if allied ontology and epistemic technologies – the contingent begins to rival the status of the social, the economic, the cultural and the political these days. A newly articulated and technoscientifically analysable aleatory materialism is beginning powerfully to emerge in which this 'reality' is rapidly becoming the focus of a new, diverse and heterogeneous, assemblage of scientifically analysable *ars combinatoria* of complex adaptive assemblages. What this regenerative biopolitics of security seeks to elicit is the generative moment from which all possible forms can be regenerated, 'the moment of emergence in which all forms are virtually infolded, *the moment of emergence considered independently of its actualizations*' (Cooper 2008: 127; emphasis added).

Here, then, the aleatory property of populations, which offered the epistemic object of early biopolitics for Foucault, becomes radically amplified and extended. It acquires an important new status. For in this newly revised biopolitical context, life in the form of Being as Becoming elevates 'the continent' into a vast new ontological and epistemic field of formation for the self-organising performativity displayed by all live manifestations of informationally driven social, cultural, political and economic power relations. Such a shift also institutes important new modes of experiencing propinquity – of how things relate, adhere and may newly belong together – in and through multi-agency, multi-medial and multi-channel processes of global dissemination. The problematisation of security, fear and danger similarly tracks this new articulation of life – begins to be thoroughly revised to the extent that it begins to revolve around this new account of the real – as it pursues the logic of such a life of becoming into the global catastrophe of the becoming-dangerous of Life to life itself. Where once Foucault interrogated the dynamics of the dangerous individual that followed from certain anatamo as well as biopolitics of life (Foucault, 2004a), for contemporary biopolitics 'being' *tout court* is becoming-dangerous.

This does not mean that subjectivities, boundaries and territorialities, which once furnished the markers of certainty that helped to demarcate and operationalise traditional security regimes, lose all significance. It means instead that subjectivities, boundaries and territory are comprehensively reconfigured as they are for example through the war on terror. They become the locale of the endless watch on the becoming-dangerous of an emergent life, of becoming haunted by the evil integral to the very reasoning faculty by which it seeks technically to govern the conduct of its conduct while appealing for that revolution in disposition without which the omnipresence of evil will remain terrorisingly omnipresent. Significant differently, they become technically determined by the preoccupation with the contingency, the complexity and the global circulation of life-like processes of every description. In particular the contingency, complexity, and circulation that co-implicate the organic and the inorganic, as well as the animal and the human, in complex ways, from the potentially turbulent flows of global capital and population migration, to foot and mouth disease, AIDS, BSC, SARS, EBOLA and migrating birds threatening the intercontinental spread of West Nile fever or Asian bird flu. Subjectivity and territoriality thus become assimilated into the problematics of global circulation in which life-like processes are bound up, such that subjects, boundaries and frontiers matter, as also do territorial authorities, but, increasingly, in as much as they provide the technologies, surveillance and self-monitoring devices for the complex governmental regulation of biopolitical global flows of every conceivable life-like formation. Technically terrorised, the moral imperative to which global liberal governance submits the governance of terror only ups the ante since the very faculty of reason upon which it relies is terrorised by the radical undecidability of the omnipresent evil of which it is integrally comprised.

It is the very instrumentality of the operational logic of life understood in this biopolitical way, the everyday positivity it demands, combined with the omnipresence of evil within the liberal account of reason to which it is also committed, and not simply the aporia of the law (Benjamin, Schmitt, Derrida and Agamben): in which there being no law that makes the law, the law is grounded in the continuous iteration of the founding violence of constituent power that makes the state of emergency the norm for this revitalised biopolitics.[2] Emergence and emergency are not only linked etymologically but also in practice – that of *species* existence in which, one might say, the norm is that immanent state of emergency in which life is understood to be a continuous process of complex, infinitely contingent and factical, circulatory transactional emergence. Biopolitical life's *raison d'être* is not to be but to become, to reproductively circulate as contingent 'species'.

To become means to be infinitely exposed, however, to the radical contingency which adaptive transactional, informationally driven, transformation without end requires. The emergency here in this new biologised, molecularised and digitalised middle ground of the real is less the suspension or force of law, then, than the norm of emergence itself. Emergence, rather than decision, is the state of emergency to which so-called information and networked societies are subject as a matter of course. Here, where there is no higher law than the law of emergence, the law of emergence finds its expression as the necessity of the radical contingency of the real as emergence. The biopolitics of security today is precisely this political emergency of emergence instituting a regime of exception grounded in the endless calibration of the infinite number of ways in which the very circulation of life threatens life, rather than some existential friend/enemy distinction. If the becoming-dangerous of being as becoming has a single counterpart to the friend/enemy distinction it is the calibration of the fitness of life for the emergency of emergence that is now said to comprise what it is to be a living being.

The terror of contingency and the contingency of terror

Terror and contingency are thus homologous, and in more ways than one. This homology arises because what is ultimately definitive of terror is not the nature of the agent or the victim. It is not the means employed or the effect sought, and neither is it simply the juridical phenomenon of being outside the laws and regular practices of war, in which distinctions between civilian and combatant are supposed to be observed, certain violent methods such as genocide forbidden and other practices, for example towards prisoners, respected. The absence of which restrictions distinguish US strategic policy, with which the British government has been shamefully complicit, in Guantanamo and elsewhere as much as they do its adversaries. Rather it is being outwith law as such. The very mode of being of terror is that radically contingent political violence, characteristic of every state of emergency including that of emergence, in which law is suspended, and the contingent necessity – the necessary contingency – of pure operationality prevails.

Irrespective of the political goals in whose name it speaks – Nationalist, Socialist, Tyrannist, Islamist, or Liberal Democratic – terror is politics conducted according to the understanding that what a politics requires in order to found or even preserve itself is not law, or even Schmittean decision, which since the advent of bodies-in-formation becoming-dangerous is deprived of its operational structure of existential enmity necessary for the friend/enemy distinction to be invoked, but the contingent instrumentality of pure operationality. Curiously, the necessary contingency and contingent necessity of pure operationality seems to make an invocation of the divine compulsory. Forget Bush and Blair for the moment. Recall

Machiavelli's footnote to Livy in *The Prince*: 'necessary wars are just wars, and when there is no other hope except in arms, they too become holy' (Machiavelli 2005: 88n).

The entire grounding of the problematic of enmity thus shifts both philosophically and politically, and a different mode of secure ordering, in fact of ordering-up the secure, comes into play. This affects codes of amity as much as it does codes of enmity, as the contingent alliance politics of opportunity and coalition warfare of the willing and the welcome proclaimed in the Pentagon QDR for 2006, for example, illustrate. For many, of course, this is the very antithesis of politics; politics reduced to the instrumentality of pure technicity, albeit such operationality now finds its current expression through the very different morphogenetic recombinatorial instrumentality of the policing through freedom of the digital and the molecular revolutions. In many respects, however, this is yet another iteration of the pure operationality required to operationalise the self-constituting political autonomy that founded the very imaginary of political modernity, not least in Machiavelli's overturning of the ancient prioritising of political from over event, privileging instead what Althusser called the aleatory materialism of evental time (Althusser 1999).

One final twist in the operational logic of liberal biopolitics emerges, here, too. Since biopolitically speaking life is understood to operate in that permanent state of immanent emergency which defines the infinite flux of its contingent emergence – we might simply say now 'the terror of its contingency and the contingency of its terror' – and notwithstanding Hans Blumenberg, a further technical resolution is sought via biopolitical governmentality to the crisis of political legitimacy of which the modern age is comprised (Blumenberg 1985). When measured against the very criteria supplied by its own operational logics – the emergency of contingent emergence – biopolitical government begins to find its nihilistic rationale and ultimate test in the operational competence it displays as a service provider of emergency relief and emergency planner of emergence. The virtual political economy of the contingent catastrophic event thus threatens to replace the virtual state of nature, that underwrote the Hobbesian protection racket, as the means by which the state power in the provision of security comes to be shared and disseminated amongst all the nodes and capillaries of network society, as it simultaneously gets comprehensively re-engineered. Witness only the institution and work of the Department of Homeland Security and the Office of Force Transformation in the Pentagon as well as the impact of Hurricane Katrina on the Bush presidency.

The terrifying inoperable force of contemporary biopolitical security politics does not therefore simply lie in the shift of the referent object of security from sovereign territoriality to life as aleatory phenomenon and the intensification of that contingency as the account of life shifts from population to complex informationally comprised adaptive system. Its true power resides in the changing understandings of the nature of what it is to be a living thing, how its changing accounts of life construe and transform the manifold of life's contingency, how the novel manifold of that contingency engenders a new manifold of reason in the neo-baroque *ars combinatoria* of risk analysis and complexity science, for example, and how all these in turn begin to constitute a newly emergent complex and dynamic field of formation for an equally complex manifold of burgeoning liberal biopolitical security discourses, governmental techniques and political rationalities characteristic of civil contingencies, strategies of national resilience and homeland security as much as they are of 'the long war' in which global liberal governance is now said to be engaged. Amongst other factors, it is the governmental political imaginary of biopolitics and the rise to prominence of biopoliticised security practices that helps account also, for example, for the many changes in security regimes, institutions and practices that have attracted so much attention since the dissolution of the Cold War: including in particular the privatisation of security, the dissolution of key demarcations between inside/outside, civil/military and public/private, and the conflation of peace and war. Once more, what is novel is not the incidence of these developments – well attested to historically – but their extension, amplification and intensification, as well as their re-coding in new biopolitical doctrines, institutions, practices and rationales: fundamentally those of the contingent circulation, connectivity and complexity that comprise the systemic interdependence of global liberal life.

Ultimately what is, therefore, at stake in biopoliticised accounts of life is the account being given of life itself; and of what it is to be 'a life' (Deleuze 2001; Agamben 1999). If you wish to contest biopolitics you cannot do so simply by taking issue with its distributive economy, geopolitical alliances, imperialising practices or murderous promotion of reproductively developmental Life planet-wide – '*they will either succeed in changing our way of life, or we will succeed in changing theirs*' (Rumsfeldt 2006, emphasis added). Ultimately, you are not only forced to contest at the level of what it is to be a living thing – in particular, of course, this living thing hitherto called '*Anthropos*' or 'Man'. Political modernity's very anthropocentrism, of which liberal geopolitics has long been a principal and principled expression, is itself now brought into question by the many ways in which its very digital and molecular revolutions have transformed what it is to be a living thing in ways that some call post-human and post-vital (Hayles 1999; Doyle 2003). There, nonetheless, in the living thing that now thinks itself beyond itself in these ways, that newly writes, speaks and kills at will as well as on command – because Brecht says it is not given to it not to kill – there nonetheless always seems more to life than meets the molecular biopolitics of contemporary biopower, just as there was always more to the human than meets the phallo-logo anthropocentric 'Man'.

As Foucault noted, biopolitics of security wager the life of the species on their own biopolitical strategies. What biopolitics of security therefore amplify in addition is how, wagering the life of the species on biopoliticising security strategies, and in the very process of biopolitically technologising life in the cause of its auto-governance, the speciated existence of biopolitical enframing finally calls itself into question whether life – whatever it is – is not to be extinguished in the name of Life.

Thus, since the vital signs of life have changed so also has what a living thing is taken to be. In the process, foundational distinctions between the organic and the inorganic, life and not life, animate and inanimate, no longer hold in the ways that they once did. Reframing inanimate material in terms of code, for example, incorporates it within the now legendary life-like capabilities of informational systems. Thus, under the regime of information, it increasingly seems as if life ought to be redefined biopolitically as 'animation'.

The implications for security politics, in particular those liberal biopolitical security discourses and practices that take life as their referent object of security, could not be more profound. The very thing that it takes as its principle of formation, Life itself becomes inoperable as what it is to be a living thing mutates in such a way that it no longer makes much sense to even pose such a question: since life-like properties can be installed in all systems by codified design, while the life-like properties of existing systems may be nullified and redesigned in virtue of the same technical capability. Here, security becomes a positive life science preoccupied with experiment and design in the fashioning of resilient self-immunising bodies rather than mere protection. Such changes provide additional reasons also why 'contingency' is so central now to the new liberal biopolitics of security, and why the very 'reason of contingency' invokes novel forms of mathematics and synthetic sciences of combination in addition to risk and probability analysis. All of these are central to how the 'long war' that now incorporates the war on terror into a war without end on behalf of life simultaneously governs through, as it is governed by, the contingency of terror. It exemplifies, precisely because it continuously threatens to purely operationalise in world destructive form, the terror of contingency. Terror piles on terror here for liberal biopolitics of security as the very principle around which it revolves – emergent Life itself – becomes inoperable as it becomes capable of negating itself.

The terror of Kantian reason

A further twist to this becoming-dangerous of Being itself is added by the Kantian philosophy that underwrites the very political reasoning of liberalism's terrorising internationalism and terrorised peace. It further clarifies how the emergency at work within it is no mere juridical emergency, Schmittean decisionism or sociological othering of others. It is integral to the reasoning of the political theology of liberalism itself.

For Kant, man does not think that he needs metaphysics. Man needs metaphysics to think. Seeking to do without metaphysics is thus irrational because reason cannot give a unified account of itself without metaphysics. This is a 'fortunate error' because it forces Man to think beyond immediate experience, instilling the desire for the beyond – which according to Kant is inherent to reason itself. Kant opens the Transcendental Dialectic, the most important section of the *Critique of Pure Reason*, with a defence of Plato's intuition. Reason is architectonic. It seeks the totality of conditions for every conditioned thing it encounters. But that totality is only given in things in themselves, to which neither reason nor understanding ultimately have access. Reason is thus impelled by its desire for knowledge beyond experience.

Reason must therefore order the concepts of the understanding through principles which in turn must be ordered through the Ideas. Such Ideas are inherently problematic since they do not arise from experience. Reason must employ the Ideas at least regulatively or it would be unable to guide understanding, and without that guide there would be no criterion of truth. If reason is to complete its architectonic task, it must adapt itself to a necessary illusion regarding the Ideas. In short, if the transcendent is a necessity, it is a necessarily regulative Idea.

Thus, in Kant's conception of moral (and political) autonomy, moral man actually has a greater need for justification and hope than the Christian does since he has become fully responsible for his actions and cannot shift responsibility back to original sin or extenuating circumstances. Neither can he conduct a proxy war against evil by stifling the inclinations of an unclean body – what Foucault called the repressive hypothesis. Instead he finds himself responsible for maintaining the correct order of incentives in his heart, but its workings are nonetheless obscure to him; in fact, in relation to evil, radically undecidable.

In Kant, the moral life is not limited to cultivating the seeds of goodness, for with them are simultaneously also planted the seeds of evil and these must be rooted out. Seen in moral terms, the life of Man for Kant is a battlefield on which the warring principles of good and evil (*das Gute, das Böse*) seek to establish dominion over human beings. In this battle no man can remain indifferent; each must choose. Morality demands 'a complete change of heart' or what Kant calls a 'revolution' in our dispositions oriented by the spirit of the moral law. After this decision, an endless struggle of self-overcoming begins 'for a humanistic holiness that no human being can achieve' (Lilla 1998: 419).

Yet what is evil? Evil is not the flesh. It must therefore be of the spirit. But it cannot not be born of original sin. Original sin is an affront to Kant's conception of moral freedom (the third postulate of practical reason). He must, therefore, explain how evil coexists with freedom and reason. He concludes that evil is a form of reasoning expressed as pride. For Kant, 'evil is a natural, if inexplicable, fact about man everywhere' (Lilla 1998: 420).

In accepting the radicality of evil along with the goodness of the inclinations Kant pushes the source of radical evil ever deeper into the heart of reasoning free will itself. Moreover according to the *Critique of Pure Reason*, we can know nothing about the will's freedom. We can only assume it. The will thus has an ineradicable propensity (*Hang*) to evil that can be activated at any moment and without it necessarily being apparent in our actions. For Kant, we can never be sure if the good or evil principle has won out. Kant calls this propensity to evil the 'foul stain on our species' and the source of our 'innate guilt' (Lilla 1998: 421). It means that since man can never fully know his own heart, 'he must become his own moral sentry, always on the *qui vive*' (Lilla 1998: 421). Evil is omnipresent and radically undecidable. That moral sentry now stands guard biopolitically over the threat that life poses to life itself in the project of promoting life.

In the beginning, God justified man. Later, Man justified God. As Man now aspires to reach his maturity man, for Kant, he must justify himself. This is not George Bush. Neither is it Tony Blair. This is Kant. Equally, however, we could say, This is not Kant. It is George Bush, and it is Tony Blair.

Kant's liberal man freed from superstitious dogmas finds he must bear alone all the burdens God once helped him carry. Man's reason is an even sterner judge. Its judgement is all the more crushing if no appeal is possible. No love for no reason operating lovingly here. In this way the technicity of biopolitics is underwritten by a widespread liberal political theology in which evil is omnipresent within the very reasoning soul of the factically finite subject condemned to govern itself as much in terms of natural rights which are not natural and freedoms which are construed as biological attributes. Subject to reason alone, the conduct of conduct becomes the liberal governance of reason terrorised by the undecidability of radical evil that is integral to its very own understanding of what it is to be governed reasonably. Being as becoming is thus doubly dangerous; dangerous because its emergent properties are undecidably contingent; dangerous also because the reason it employs to govern the terrorising contingency of its life of emergence is itself terrorised by the undecidability of the radical evil of which its reasoning faculty is said to be also comprised (Copjec 1996; Neiman 2002; Bernstein 2002).

Liberal self-becoming thus lives in a double jeopardy; practical fear of the contingency of emergence and moral fear of failing in the pursuit of self-perfection in the presence of omnipresent evil. If the moral law that moves him by instilling in him humility also reminds him, according to Kant, that he will never be humble enough, and that he alone bears the responsibility for this, the practical law that also now moves him instils the terror that his techniques will fail the test of contingent emergence as well. This creature,

> lives in fear and trembling no less than the Christian believer. Perhaps more so since what he fears is also himself ... There is no redeemer; we must redeem ourselves. Nor can our redemption take place through a single act of the will. After the decision for morality, the passion of autonomous moral man has just begun. For him everyday is a Calvary.
>
> (Lilla 1998: 421–2)

Lilla drives home the point that Kant's political theory is dependent upon the *homo post-Christianus* which Kant develops especially in *Religion within the Bounds of Mere Reason* (1998) and *The Conflict of the Faculties* (1979). He adds that,

> The rebirth of Christianity out of the spirit of the Enlightenment was surely not what Kant had in mind ... What is most striking about his theological-political tracts is how far Kant himself actually advances in this direction. Moreover it is not an altogether liberal direction.
>
> (Lilla 1998: 426)

Part of the argument here, however, is that the liberal direction itself is not altogether 'liberal'. It requires a commonwealth of believers whose dogmatism is as corrosive of its own accounts of freedom as its policing is corrosive of its free institutions.

For, considered politically, the dynamics of the state of nature teach us the necessity of coercive public laws. Considered ethically, however, a state of nature exists whenever individuals, even within a political state, are not members of the same 'church', that is do not recognise a common authority over morality. Instituting such a commonwealth creates a violently divided loyalty: that between local political communities and this global civil community. It is a divided loyalty similar to that which once vexed the division of Church and State. Tony Blair, for example, has been committed to such a doctrine since 1999:

> So, for me, before September 11th, I was already reaching for a different philosophy in international relations from a traditional one that has held sway since the treaty of Westphalia in 1648; namely that a country's internal affairs are for it and you don't interfere unless it threatens you, or breaches a treaty, or triggers an obligation of alliance.
>
> (Blair 1999)

This means that even if the public peace is maintained through national and international law, including for example that of non-intervention in the affairs of other states, the moral state of man will decline as social and political interaction transform man's naturally good inclinations into wicked ones. However, since in Kant's view the enforcement of private morality is not a public political matter – morality is by definition for him non-coercive – a different form of community must be instituted to guide it. This ethical commonwealth as he calls it (*ethisches gemeines Wesen*) is not political. It is not government, at least in the public sense, and it is not sovereign, but it is cosmopolitan. It promotes the good principle and reduces the likelihood of our exciting the evil principle in each other. The ethical commonwealth is that cosmopolitan community, transcending both Church and State, tasked in the Kantian scheme with the moral improvement of Man and against whose moral standards his political conduct must be measured. The historical mission of this ethical civil commonwealth transcends even that of international law. For while international law can establish cosmopolitan relations between states under differing political laws, 'only the laws of the ethical commonwealth embrace the whole of humanity' (Lilla 1998: 428)

Limiting religion to inner morality Kant struck a blow against Christianity. By elevating morality to the ultimate aim of human activity, indeed of creation itself, Kant nonetheless simultaneously also promoted the post-religious Christian Church into the historical vanguard of man's moral reform and political enlightenment. Reformed religion makes its contribution to the work of this vanguard directly through contribution to the supra-political ideal of an ethical, civil and cosmopolitan, commonwealth. Here in this liberal ideal of progress towards peace

Christianity and modernity are not antagonists but partners in a reworked onto-theologically promoting baroque drama. If we were Greeks we might say tragedy. But we are not Greeks. They were as foreign to themselves as they are to us, and as we are to ourselves, but differently.

While that drama has gone biopolitical, and the life of its biopolitics has become recombinatory emergence, the nature of the drama remains that of *Trauerspiele* whose nihilism is amplified and intensified by the ways in which it receives expression in contemporary molecular biological terms. It is no exaggeration to say that this is precisely what has Bush and Blair on the cross; acting out the liberal mourning play of power in their respective Christian idioms (Benjamin 1998).

The erasure of 'man' and the targeting of life

> '*techne* loves *tyche* and *tyche* loves *techne*.'
>
> (Agathon)[3]

Operationally speaking, this biopoliticised state of continuously emergent emergency is one in which the West's foundational preoccupation with species as well as state security also goes hyperbolic. For Life here is, of course, the life of 'species existence'. Such hypersecurity is no simple politics of fear despite the manifold of contingent fears that fuel it. Biopolitically speaking, hypersecurity is integral to the operational logic that is definitive of the very Life of contemporary biopolitics. It thus exceeds, even as it incorporates, the traditional security problematic of the friend/enemy distinction. For, under the biopolitical regime of emergent emergency, it is no longer adequate to judge life-like bodies in terms of the essence of that existential otherness definitive of the 'enemy' alone, for every body is a continuously emergent body in formation comprised of contingently adaptive rather than fixed properties.

Such 'bodies-in-formation' render the friend/enemy distinction inoperable in the same way that the Life of pure operationality renders the state of emergency normal. In fact, these days, since every body is informationally defined as emergent in terms of the interaction of codes that make it up, every (informational) body is quite literally now understood to be a body-in-formation. The operational logic of the emergent life of contemporary biopolitics lived as an immanent life of continuous contingent emergence, thus institutes a life of becoming whose hypersecurity politics is fundamentally a politics of dangerous becoming. Herein also lies what Derrida calls its auto-immunising terror, in which the very processes involved in the protection of life are destroyed in its very securing (Derrida 2005).

It is not what a body is that makes it biopolitically a threat, then, but what a body might potentially become. Pluripotent, we simply do not know, because we have not yet seen an end, to what a body – of any description: individual, collective, cellular or machinic – might become. Hence, the hypersecurity of becoming-dangerous. The locus of what a body might become, in this age of recombinant biopolitics, is, however, now said to lie in the dynamic biometry of its code. Expressing the very operational logic of biopolitical life, it is no wonder that biometrics is fast becoming one of the techniques of choice for biopolitical securing. Securing bodies-information ineluctably becomes a process of securing the codes of which they are comprised. Little wonder, either, that an Alaskan businessman, John Moore, was disturbed one day to discover that the University of California at Los Angeles had patented his body parts and licensed them to the Sandoz Pharmaceutical Corporation (Norris 2000: 1).

Here, then, the new preoccupation with terror while nonetheless still regularly articulated in geopolitical terms – of rogue states, failed states, extremist regimes and the spread of weapons of mass destruction – has very largely also been assimilated into this ontologically and epistemically different politics of life itself that has brought global liberal governance to the point of moral dissolution through the military strategic calamity by which it is currently pursuing its martial vision of the free life it espouses; such that even to be poor is to be terrifyingly dangerous, *in potentia* if not in actuality. If trashing the poor for the threat that they pose to freedom is an old game, it is being played out more remorselessly now according to new rules and with a ruthlessness that comes from no longer having NGOs speaking for them; since the NGO community has long been incorporated into the vitalistic ideology of power/knowledge that oppresses them.

It is through the remorseless biopolitical assay of life that the poor are rendered terrifyingly dangerous to life. Biopolitics simply lives for its obsession with the audit of existence. For the continuous assay of life is necessary to specify the very eligibility to life as well as the eligibilities which life biopolitically accords to life. How would biopolitics know how to promote and enhance life if it did not constantly take the measure of life? And what is it to do, when constantly taking the measure of life, if it discovers life intractable to improvement, or even inimical to life itself? It must specify correction and administer punishment. In the final event it must also equip itself to say who shall live and who shall die in the name of life itself. A continuously recidivist form of life traditionally identified by class, racial, sexual and gendered markers, poverty becomes poor in species being. Poor in species, however, transcends poverty. It indicates all those insurgent against the biopolitical regime of life; check out the bio-pics of Islamist terrorists drawn from sophisticated backgrounds, the characterisation of political regimes as rogue and the generic fear of 'extremists' licensed by QDR 2006 for extermination.

The deserving poor were always given a break. Without them the moral order of improvement could not function. All life, however, is now caught in the semantic web of biopolitics. Ineligibilities for life proliferate in its ruthless promotion of life. Indigence thus mutates biopolitically into 'poor in species being'. All are capable of this. Notwithstanding the persistence of old codings none may now appeal against it to the privilege of class, race, sex or gender – benefit of clergy was lost long ago in a different kind of political upheaval. In the name of life's enhancement, war re-enters the biopolitical metrics of power/knowledge over life with a vengeance against ineligible or inimical life threatening in one degree or another the promotion of life as species being.

Conclusion

I want to return by way of conclusion to the relation of biopolitics to spatiality and temporality as well as to power. Whereas under sovereign juridical regimes of power, monopolising the taking of life helped demarcate the territorial integrity of the sovereign whose power was infinite, under biopolitics the very evental 'taking place' of life now becomes the locus of a power for which infinite flux of immanent contingent change is central. (Britain's former Prime Minister Tony Blair insists on calling this 'modernisation'.) I am not arguing that all this is a matter of sheer luck or pure chance. Once more it has to be emphasised that contingency has a history, and that the history of modern liberal biopolitical contingency is thereby characterised by its own reason as well. Paradigmatically that matrix of reason includes capitalism, but it also exceeds as it re-informs capitalism's pure economic calculus of circulation – the advent and exponential growth of the futures markets is a case in point. Here what Marx after Aristotle called the '*perpetuum mobile* of circulation' (Marx 1976: 226), which defines capitalism, is also embedded within a wider *perpetuum mobile* of biopolitical reproduction, where the innermost principle of Life itself is survival as adaptive contingency. In respect of spatialisation one might also say, iconically, that Mercator projection cedes political priority, if not of course all political and economic significance, to the new life-spatialising sciences such as those for example of genetic mapping, virtual cartography and epidemiology (Rogers 1996, 2000, 2002, 2004; McNally 2005). (Our biological imagination is governed by the metaphors of mapping as much as it is by the metaphors of information and code [Gaudillière and Rheinberger, 2004].) New tabulations of life, as contingent adaptation whose very circulation amplifies and intensifies all the systemic hazards, risks, dangers, pathologies and epidemics to which it is subject, engender a new space and time of biopolitical existence for the twenty-first century. Hence the widespread medicalisation of security discourse and practices – from asymptomatically ill beings and preventative medicine to asymptomatically dangerous beings and preventative war. Hence also the securitisation of medicine as integral part of the strategy of national resilience for dealing with catastrophic event and terroristic attack.

When at the end of *The Order of Things* Foucault imagined that 'one can certainly wager that man would be erased, like a face drawn in the sand at the edge of the sea' (Foucault 1989: 387) we might also observe how the quasi-transcendentals of Labour, Life and Language that he identified as demarcating 'Man' have been mutating into Information, Animation and Code. In the process, Circulation, Connectivity and Complexity offer replacements. The new sciences of Circulation – specifying a new terrain of value across which life as such, and not simply 'Man' as a living being, is now ordered – offer different accounts of dissemination, proliferation and propagation, especially of how new threats and dangers are constituted and very rapidly amplified by the very same systems that circulate life. The new sciences of Connectivity give novel accounts of the global–local propinquity, adhesion, adherences, proximities, associations, alliances, virtualities, realities and belonging that constitute life. The new Complexity sciences study the life-like properties of complex adaptive systems and offer novel accounts of the spontaneous non-linear phase changes that are now said to constitute the vital signs of such life's very animation. What lends unity to the field of new knowledge demarcated by this new triangulation of quasi-transcendentals is no longer 'Man', and in a sense, also, not even Life but 'Contingency'. Thus has the modern been morphing into a new hypersecuring liberal biopolitical positivity in which – the 'cult of man' erasing itself biopolitically – the odds on species extinction continue, in lethal paradox, to shorten remorselessly.

Notes

1 Some traditional geopolitical analysts argue that the current situation stems from a failure to practise geo-strategy rather than the reverse (Strachan 2005). Strachan's chapter nonethless begs a whole metaphysics of the subject which is why its traditional teleological account of strategisation is itself profoundly problematic.

2 One must add that for Agamben, as well as for Derrida albeit differently, the aporia of the Law is the aporia of Language as well. Indeed in Agamben's messianic account of the limit situation Law and Language are one (Agamben 2005). For the profound difference between Agamben and Derrida, see Thurschwell (2005).

3 *Tyche* is the ancient Greek word for luck or chance. This quotation is from Vatter (2000: 41). Reflections on *tyche* and its correlate intelligence *metis* may be found in Vernant and Vidal-Naquet (1990), Detienne and Vernant (1978), Pucci (1985) and Dillon (2002a). Vatter points out that before Plato, Greek thought treats event either as a more or less immediate manifestation of the divine (e.g. Hesiod) or as a matter of chance (Anaximander). In the first instance, providence is strictly a religious affair; in the latter case, chance is strictly a political affair, often used to provide a critique of religion. There is, he says, no attempt to unite religion and politics into a political theology or a theologically founded politics as one finds in Plato's *Laws*. Thus Plato is the first to articulate the project of onto-theology: thus, onto-theologically, 'with the *Laws* the first form of political theology is born' (Vatter 2000: 42).

Bibliography

Agamben, Giorgio (1999) 'Absolute Immanence', in Agamben, *Potentialities* (Stanford, CA: Stanford University Press), pp. 220–40.

Agamben, Giorgio (2005) *The Time that Remains: A Commentary on the Letter to the Romans* (Stanford, CA: Stanford University Press).

Althusser, Louis (1999) *Machiavelli and Us* (London and New York: Verso).

Althusser, Louis (2006) *Philosophy of the Encounter: Later Writings, 1978–87* (London and New York: Verso).

Aristotle (1986a) *De anima. Aristotle in Twenty-Three Volumes* (Cambridge, MA: Harvard University Press).

Aristotle (1986b) *On The Soul; Parva Naturalis; On Breath* (Cambridge, MA: Harvard University Press).

Arquilla, John and Ronfeldt, David F. (2000) *Swarming and the Future of Conflict* (Santa Monica, CA: RAND).

Arquilla, John and Ronfeldt, David (2003) 'Swarming – The Next Face of Battle', *Aviation Week & Space Technology*, 29 September. <www.rand.org/blog/2003/09/swarming----the-next-face-of-battle.html>.

Beck, Ulrich (1992) *Risk Society: Towards a New Modernity* (Newbury Park, CA: Sage Publications).

Benjamin, Walter (1996) 'Critique of Violence', in Benjamin, *Selected Writings*, ed. Marcus Bullock and Michael Jennings (Cambridge, MA: Belknap Press), vol. 1, pp. 236–52.

Benjamin, Walter (1998) *The Origin of German Tragic Drama* (London and New York: Verso).

Bernstein, Peter L. (1998) *Against the Gods: The Remarkable Story of Risk* (New York: John Wiley).

Bernstein, Richard J. (2002) *Radical Evil: A Philosophical Interrogation* (Cambridge: Polity Press).

Blair, T. (1999) Speech to the Chicago Economic Club, 22 April.

Blumenberg, Hans (1985) *The Legitimacy of the Modern Age* (Cambridge, MA: MIT Press).

Bonta, Mark and Protevi, John (2004) *Deleuze and Geophilosophy: A Guide and Glossary* (Edinburgh: Edinburgh University Press).

Cooper, Melinda (2008) *Life as Surplus: Biotechnology and Capitalism in the Neoliberal Era* (Seattle: University of Washington Press).

Copjec, Joan (ed.) (1996) *Radical Evil* (London and New York: Verso).

Daston, Lorraine (1988) *Classical Probability in the Enlightenment* (Princeton, NJ: Princeton University Press).

Deleuze, Gilles (2001) 'Immanence: A Life...', in Deleuze, *Pure Immanence: Essays on a Life* (New York: Zone Books), pp. 25–33.

Derrida, Jacques (2005) *Rogues: Two Essays on Reason* (Stanford, CA: Stanford University Press).

Detienne, Marcel and Vernant, Jean-Pierre (1978) *Cunning Intelligence in Greek Culture and Society* (Brighton: Harvester Press).

Dillon, Michael (2002a) 'Intelligence Incarnate: Martial Corporeality in the Digital Age', *Body and Society*, 9(4): 123–47.

Dillon, Michael (2002b) 'Network Society, Network-Centric Warfare and the State of Emergency', *Theory, Culture & Society*, 19(4): 71–9.

Dillon, Michael (2005) 'Global Security in the 21st Century: Circulation, Complexity and Contingency'. ISP/NSC Briefing Chapter.

Dillon, Michael (2007) 'Governing through Contingency: The Security of Biopolitical Governance', *Political Geography*, 26(1): 41–7.

Dillon, Michael and Lobo-Guerrero (2006) 'The Biopolitical Imaginary of Species Being and the Freedom to Underwrite in the Molecular Age' (Lancaster University).

Doyle, Richard (1997) *On Beyond Living: Rhetorical Transformations of the Life Sciences* (Stanford, CA: Stanford University Press).

Doyle, Richard (2003) *Wetwares: Experiments in Postvital Living* (Minneapolis: University of Minnesota Press).

Edwards, Sean J. A. (2000) *Swarming on the Battlefield: Past, Present, and Future* (Santa Monica, CA: RAND).

Ericson, Richard and Doyle, Aaron (2004) 'Catastrophe Risk, Insurance and Terrorism', *Economy and Society*, 33(2): 135–73.

Foucault, Michel (1989) *The Order of Things: An Archaeology of the Human Sciences* (London and New York: Routledge).

Foucault, Michel (2003) *Society Must Be Defended: Lectures at the Collège de France, 1975–76* (London: Picador).
Foucault, Michel (2004a) *Abnormal: Lectures at the Collège de France, 1974–1975* (New York: Picador).
Foucault, Michel (2004b) *Naissance De La Biopolitique: Cours au Collège de France, 1978–1979* (Paris: Gallimard/Seuil).
Foucault, Michel (2004c) *Sécurité, Territoire, Population: Cours au Collège de France, 1977–1978* (Paris: Seuil/Gallimard).
Foucault, Michel (2007) *Security, Territory, Population: Lectures at the Collège de France, 1977–78*, ed. Michel Senellart (Basingstoke: Palgrave Macmillan).
Gaudillière, Jean-Paul and Rheinberger, Hans-Jörg (eds) (2004) *From Molecular Genetics to Genomics: The Mapping Cultures of Twentieth-Century Genetics* (London and New York: Routledge).
Gell-Mann, Murray (1994) *The Quark and the Jaguar: Adventures in the Simple and the Complex* (New York: W. H. Freeman).
Gleick, James (1987) *Chaos: Making a New Science* (London: Sphere Books).
Hacking, Ian (1982) 'Biopower and the Avalanche of Printed Numbers', *Humanities in Society*, 5: 279–95.
Hacking, Ian (1990) *The Taming of Chance* (Cambridge: Cambridge University Press).
Hacking, Ian (1995) *The Emergence of Probability: A Philosophical Study of Early Ideas about Probability, Induction and Statistical Inference* (Cambridge: Cambridge University Press).
Hayles, Katherine (1999) *How We Became Posthuman: Virtual Bodies in Cybernetics, Literature, and Informatics* (Chicago, IL: University of Chicago Press).
Introna, Lucas D. (2005) 'Disclosive Ethics and Information Technology: Disclosing Facial Recognition Systems', *Ethics and Information Technology*, 7: 75–86.
Introna, Lucas D. and Wood, David (2004) 'Picturing Algorithmic Surveillance: The Politics of Facial Recognition Systems', *Surveillance & Society*, 2(2–3): 177–98.
Kant, Immanuel (1979) *The Conflict of the Faculties* (Lincoln, NE: University of Nebraska Press).
Kant, Immanuel (1998) *Religion within the Boundaries of Mere Reason and Other Writings*, ed. Allen W. Wood and George Di Giovanni (Cambridge: Cambridge University Press).
Kay, Lily (1993) *The Molecular Vision of Life: Caltech, the Rockefeller Foundation, and the Rise of the New Biology* (Oxford: Oxford University Press).
Kay, Lily (2000) *Who Wrote the Book of Life? A History of the Genetic Code* (Stanford, CA: Stanford University Press).
Kellert, Stephen H. (1993) *In the Wake of Chaos: Unpredictable Order in Dynamical Systems* (Chicago, IL: University of Chicago Press).
Knorr-Cetina, Karin (1999) *Epistemic Cultures: How the Sciences Make Knowledge* (Cambridge, MA: Harvard University Press).
Krüger, Lorenz, Daston, Lorraine, Heidelberger, Michael, Gigerenzer, Gerd, and Morgan, Mary S. (eds) (1987) *The Probabilistic Revolution* (Cambridge, MA: MIT Press).
Landa, Manuel de (2002) Intensive Science and Virtual Philosophy (London: Continuum).
Lennox, James G. (2001) *Aristotle's Philosophy of Biology: Studies in the Origins of Life Science* (Cambridge: Cambridge University Press).
Lilla, Mark (1998) 'Kant's Theological-Political Revolution', *Review of Metaphysics*, 52(2): 397–434.
Machiavelli, Niccolò (2005) *The Prince* (Oxford: Oxford University Press).
MacKay, Donald MacCrimmon (1969) *Information, Mechanism and Meaning* (Cambridge, MA: MIT Press).

McNaghten, Phil, Grove-White, Robin, and Wynne, Brian (2004) 'Nanotechnology, Risk and Sustainability: Moving Public Engagement Upstream'. ESRC Sustainable Technologies Programme, Project Progress Report.

McNally, Ruth (2005) 'Sociomics! Using the Issue Crawler to Map, Monitor and Engage with the global proteomics research network', *Proteomics*, 5: 3010–16.

Marx, Karl (1976) *Capital: A Critique of Political Economy* (Harmondsworth: Penguin).

Neiman, Susan (2002) *Evil in Modern Thought: An Alternative History of Philosophy* (Princeton, NJ: Princeton University Press).

Norris, Andrew (2000) 'Giorgio Agamben and the Politics of the Living Dead', *Diacritics*, 30(4): 35–58.

Norris, Clive and Armstrong, Gary (1999) *The Maximum Surveillance Society: The Rise of CCTV* (Oxford and New York: Berg).

Novas, Carlos and Rose, Nikolas (2000) 'Genetic Risk and the Birth of the Somatic Individual', *Economy and Society*, 29(4): 485–513.

Omand, Sir David (2005) 'National Resilience Priorities for UK Government'. <www.rusi.org>.

Protevi, John (2001) *Political Physics: Deleuze, Derrida, and the Body Politic* (London and New York: Athlone Press).

Pucci, Pietro (1985) *Odysseus Polutropos: Intertextual Readings in 'The Odyssey' and 'The Iliad'* (Ithaca, NY: Cornell University Press).

Rheinberger, Hans-Jörg (1997) *Toward a History of Epistemic Things: Synthesizing Proteins in the Test Tube* (Stanford, CA: Stanford University Press).

Rogers, R. (1996) 'The Future of STS on the Web', *EASST Review*, 15.

Rogers, R. (2002) 'The Issue Crawler: The Makings of Live Social Science on the Web', *EASST Review*, 31.

Rogers, Richard (2000) *Preferred Placement: Knowledge Politics on the Web* (Maastricht: Jan van Eyck).

Rogers, Richard (2004) *Information Politics on the Web* (Cambridge, MA: MIT Press).

Rose, Nikolas (1999) *Governing the Soul: The Shaping of the Private Self* (London: Free Association Books).

Rose, Nikolas (2001) 'The Politics of Life Itself', *Theory, Culture & Society*, 18(6): 1–30.

Rose, Nikolas (2003) 'Neurochemical Selves', *Society*, 41(1): 46–59.

Rumsfeldt, Donald (2006) You Tube. <https://www.youtube.com/watch?v=GiPe1OiKQuk>, last accessed 14 October 2014.

Sebeok, Thomas (1999) 'The Sign Science and the Life Science', *Applied Semiotics / Sémitique appliquée*, 3: 85–96.

Sebeok, Thomas (2001) *Global Semiotics* (Bloomington: Indiana University Press).

Strachan, Hugh (2005) 'The Lost Meaning of Strategy', *Survival: Global Politics and Strategy*, 47(3): 33–54.

Strauss, Jonathan (1998) *Subjects of Terror: Nerval, Hegel, and the Modern Self* (Stanford, CA: Stanford University Press).

Taleb, Nassim Nicholas (2001) *Fooled by Randomness: The Hidden Role of Chance in the Markets and in Life* (New York: Texere).

Thacker, Eugene (2003) *Biomedia* (Minneapolis: University of Minnesota Press).

Thurschwell, Adam (2005) 'Cutting the Branches for Akiba: Agamben's Critique of Derrida', in Andrew Norris (ed.), *Politics, Metaphysics, and Death: Essays on Giorgio Agamben's Homo Sacer* (Durham, NC: Duke University Press), pp. 173–97.

Vatter, Miguel E. (2000) Between Form and Event: Machiavelli's Theory of Political Freedom (Dordrecht: Kluwer Academic Publishers).

Vernant, Jeanne-Pierre and Vidal-Naquet, Pierre (1988) *Myth and Tragedy in Ancient Greece* (New York: Zone Books).

CONCLUSION

Acceleration of the baroque katechon

> Instead whereof the Greeks have prosopon, which signifies the Face, Persona in Latin signifies the disguise, or outward appearance of a man, counterfeited on the stage; and sometimes more particularly that part of it, which disguiseth the face, as a mask, or vizard [visor]: and from the stage, hath been translated to any representer of speech and action, as well in tribunals, as theatres.
>
> (Hobbes 1960)

The living dead

> I don't claim to kill others with my writing. I only write on the basis of the other's already present death.
>
> (Foucault 2013: 42)

In 1968 Foucault agreed to a series of interviews with the critic Claude Bonnefoy. This project was soon abandoned, but a transcript of the initial interview survived and has recently been published in English (Foucault 2013). Foucault said many things about his work, many of which were inconsistent and deliberately designed to avoid being labelled. What he said to Bonnefoy did not, therefore, define him. He was, in any event, beyond definition. But it does prompt this final reflection on the political analytic of finitude, and on how Man and Life helped give concrete political form to factical finitude.

'For me,' Foucault said, 'writing means having to deal with the death of others, but it basically means having to deal with others to the extent that they're already dead … I'm in the situation of the anatomist who performs an autopsy' (2013: 40–1). 'I also understand,' he said, 'why people experience my writing as a form of aggression'

(2013: 41). 'They feel there is something in it that condemns them to death' (2013: 41). He clarifies: 'In fact, I am much more naïve than that. I don't condemn them to death. I simply assume they're already dead. That's why I am so surprised when I hear them cry out. I'm as astonished as the anatomist who became suddenly aware that the man on whom he was intending to demonstrate has woken up beneath his scalpel. Suddenly his eyes open, his mouth starts to scream, his body twists, and the anatomist expresses his shock: "Hey, he wasn't dead!"' (2013: 41). He then explains that this, so-called, 'postmodern irony' is, in fact, an expression of his astonishment that the already presumed dead scream out in the middle of these living autopsies. In short, he says, 'I don't claim to kill others with my writing. *I only write on the basis of the other's already present death*' (2013: 42, emphasis added).

I take this to mean – it may not, so I claim usufruct – that the most aggressive autopsies that Foucault conducted were on the living dead of Man and Life, the very figures through which modern subjectivity is lived out. To the degree that modern subjects experience Foucault's writing as a form of aggression, therefore, it is precisely because his autopsies threaten the very investment they have made in their own processes of subjectification. For, Man and Life do not simply live their 'already present death(s)' figuratively, as if the figurative were not itself one of the most potent forms of materialisation. Like gods, they walk the earth. Via new figurations and correlations of the life–death nexus, they have provided the very strategic figures around which the political morphology of baroque and neo-baroque politics, and the epistemic morphology of baroque science and neo-baroque technoscience, have combined to produce the political modernisation and power relations that we now enjoy. The lives of the self-governing subjects of such modernisation are, thereby, increasingly fated, in the name of security, to living a voluntary servitude to the indefinite governance of the finite things and processes to which they are reduced under the strategic regimes of both Man (geopolitics) and Life (biopolitics). Somewhere in this game, life's enigma unfolds its other expressions, fugitively and in defiance of the katechontic securitisation to which much modern politics has itself been reduced. But this is another story.

The baroque

The societies of the North Atlantic basin have never been so secure. That security is a function of such a wide variety of security institutions and practices it seem pointless to name them, for they penetrate into the very weft and warp of everyday life. Pointless to name them, but not pointless to re-describe them by giving them a different name. They are in every respect baroque.

Ordinarily addressed as a period and or a style of art, an aesthetic, the baroque names, instead, a space of problematisation, and a mode of operationalisation. If factical finitude is the condition of possibility for the baroque, as well as for modern politics, the baroque names factical finitude's condition of operability. No more so, in fact, than in respect of the conditions of operability of modern politics of security.

What is being secured, in the securing of Life whose very vital signs are now construed as generically dangerous to itself, or of the sovereign that is fated to fail the standards of sovereign being, is the continuous suspension of any other expression of politics. It is a suppression that works through substitution, the substitution of repeated clichés, marvels and spectacles concerning the 'now', the 'future', 'potency', 'radical contingency' and survivability in a world whose very security apparatuses have turned politically motivated killing into an industrial and commercial processes integral to our civilization but capable also of threatening the survivability of planetary life. At the beginning of the twenty-first century, these now seek their expression in tropes that intensify security's baroque fixation with morbidity, mortality and modernity: complex adaptation and change, competitiveness, emergence, catastrophic emergency, the event that saves, the event that threatens, and, above all, resilience – the capacity to endure by becoming something else. To survive is to submit, through holocaust or social therapy, to self-annihilation as governmental necessity. We might therefore envisage the baroque character of our politics of security as the attempt to unify these themes within synthetic narratives of competitiveness, identity, the enemy, or simply terror. The past of security and war is forgotten by continuous repetition in spectacle, simulation and practice. These are now organised through scopic regimes employing different orders of signification and representation as well as novel technologies.

To give our contemporary politics of security another name is, therefore, a deliberate device. It gives them another face. In rhetoric this manoeuvre is called *prosopopoeia*. Strictly speaking, *prosopopoeia* gives a face to that which does not have a face. Our security politics are preoccupied, however, with presenting a face that will move the world, a face for the world to accept at face value. Having a face is, therefore, not the issue here. The issue is what face. I have attempted to give our politics of security another face, one capable of refracting its face of baroque display. *Prosopopoeia* is a difficult art. It does not seek to achieve closure or finality as some forms of narrative, history and positivity do. It does not labour under the rule of verisimilar adequacy, and it does not suborn itself to policy or governmental relevance. It seeks to represent not that which is absent, 'but that whose presence is so intense that we can only feel it and … see it from a safe distance' (Godzich and Spadaccini, in Maravall 1986: xiii–xiv).

Initially, the political programme of the baroque was the formal answer of the monarchical-seigniorial segments of sixteenth-century Spanish society, to the assaults launched against the traditional étatist structure taking shape in early modern Spain (Godzich and Spadaccini, in Maravall 1986: xvii). One of its most distinguished historians, José Antonio Maravall, defines the baroque, 'as a culture provoked by a cultural crisis of major proportions, one that was felt in all of Europe, and perhaps most intensely in Spain, during the greater part of the seventeenth century' (Godzich and Spadaccini, in Maravall 1986: xviii). Where, once, baroque politics included defence of the monarchy, the safeguard of honour as the *raison d'état* of individual and social life, and the constant reaffirmation of love as a universal justification (Godzich and Spadaccini, in Maravall 1986: xix), our baroque politics of security is a vast palimpsest of rules of truth and truths of rule, constantly reaffirming peace, rights and justice as universal justifications for security and war. The baroque names its mode of operativity, rather than a period.

I have extended this description of it to the generic crisis induced by factical finitude and the changing baroque politics of security that helps distinguish the states and societies of the North Atlantic basin at the beginning of the twenty-first century. As with the early baroque, so also with the twenty-first century we are confronted with a political culture directed towards the multitude of anonymous and, therefore, potentially disruptive individuals – homegrown 'terrorists', for example – no longer simply concentrated in the cities, but capable of circulating globally. Its mode of operativity is distinguished by the effort that is spent on casting the political subject as political spectator, whose voluntary servitude to indefinite governance is secured through the spectacles of security, catastrophic emergency and war, as much as through their everyday governmentalisation.

Deleuze noted, acutely, that, '[t]he essence of the baroque entails neither falling into nor emerging from illusion but rather realizing something in illusion itself' (Deleuze 1992: 124). A fuller depiction of the neo-baroque character of contemporary security politics, therefore, requires further studies. Among other things, these would have to revisit the problematization not only of spectacle and the virtual, but all modes of making manifest politically and governmentally in which these are intimately involved as a matter of priority. But this has to wait, and for the moment I can only add a coda to the present book, by gesturing towards how politics of security in the twenty-first century are distinguished by the ways in which they exceed the katechontic security politics of the baroque sovereign that I introduced, through Reinhart Koselleck and Carl Schmitt, in the first chapter of the book.

Politics of security remain katechontic, fated to pursue the infinite deferral of the very finitude of which they are comprised, and from which they also take their warrant to conduct the infinite securitisation of finite things and processes of becoming finite. But this katechontic enterprise changes its character as we not only move from addressing modern sovereign geopolitics to biopolitics, but also as we move from the baroque traits of the early modern era to those of the neo-baroque of our own times. What is additionally interesting is that, as the modern geopolitics of the West has become a largely biopolitical enterprise, so also has the katechontic task of restraint become acceleration of the very forces that the katechon was once said to restrain – chaos, lawlessness and anarchy, since order now is commonly said to arise from chaos. Katechontic politics of security at the beginning of the twenty-first century thus aim to become the very anomic chaos originally stigmatised in the anarchy problematic of classical international relations theory, which so extolled the necessity of the sovereign state, and of statecraft, simultaneously also fostering the belief that there was no stagecraft to statecraft, in explicitly katechontic terms. No longer simply committed to restraining the coming of the end, the katechontic enterprise of the biopolitics of security of the twenty-first century is now much less committed to restraint than to acceleration, acceleration of the vital forces of being-in-formation and becoming-dangerous that exceed finitude, thereby offering the prospect of securely commanding it.

This, then, is less a matter of resisting the end, than cultivating the vitality whose power is not presence, but the power to make present. Contra Schmitt, this is a less a matter of the *trompe l'oeil* of 'the miracle', a baroque trope if ever there was one, than of the novel powers of artifice and illusion realised through digital powers of the sign. As Deleuze observed, it is, of course, not the illusion that counts but what is realised through illusion. As one, now infamous, aide to George W. Bush blankly put it. 'We are an empire now, and when we act, we create our own reality. And while you're studying that reality – judiciously, if you will – we'll act again, creating other new realities, which you can study too, and that's how things will sort out. We're history's actors … and you, all of you will be left to study what we do' (quoted in Egginton 2010: 1). Except, of course, as Benedetto Croce long ago observed, in the epigraph that heads this book, they do not, in fact, know what they do. It is not as if this were not, however, an all too familiar world – a madness that occurs in the full light of day.

The katechon

> I believe in the Katechon; for me it is the only possibility for understanding history and considering it meaningfully as Christian.
>
> (Carl Schmitt, quoted in Bradley 2014: 1)

Schmitt's declaration of faith in the katechon is deeply revealing for its omissions. But it is Schmitt's little noted commission, imparting a powerful new twist to the katechon, whose charge can also be directly observed in contemporary politics of security, that is perhaps even more revealing (Dillon and Reid 2009). In any event, Schmitt's omissions combine with this commission not only to confirm the overwhelming baroque character of the politics of security of factical finitude in the twenty-first century, but also to signify what distinguishes it.

How do the omissions signify the baroque character of modern politics of security? Well, there is, for example, no mention of the Christian eschaton, the promised return of the Messiah, divine love, divine forgiveness and the prospect of life everlasting. The Christian katechon – such as it is – cannot be divorced from this wider semantic field of religious reference and faith formation. But the history of redemption, in which the katechon played a recurrent if problematic part, seems reduced by Schmitt to the history of the katechon alone, indeed to political, or what he called world history as katechontic.

Schmitt's omission is, then, the Christian God itself. Or, at least, if God is not entirely omitted, he has become that *deus absconditus* whose withdrawal from the world, via Christian theology among other things, precipitating the advent of factical finitude, nonetheless left its trace on the world, through the strategic figure of Man. Schmitt's world is the world of Koselleck's sovereign. And the world of Koselleck's sovereign is the world of the baroque. Schmitt's *deus absconditus* thus gives us the first clue to the baroque character of the modern katechontic sovereignty theorised by Schmitt. The historical baroque, its very aesthetic morphology, is known for the theatrical tropes that it puts into play. These include, anamorphosis, *mise en abyme*, *trompe l'oeil*, the juxtaposition of disparate terms, the proliferation of *décor* and the conscious embrace of artifice (Egginton 2010). It is a world of appearance and spectacle in which: 'the organizing logic … is a theatrical one in which the space of representation is severed onto a screen of appearances and the truth presumed to reside behind it, and it is this basic problem of thought that underlies the multiple strategies that baroque aesthetic production puts into play' (Egginton 2010: 127). These fascinated Schmitt as well, although he has little to say about them explicitly, or the machination and intrigue of court politics that necessarily surrounds the sovereign, except in his observations on 'acclamation' (Dean 2014). For Egginton, and other contemporary analysts, the baroque aesthetic arises out of wider historical conditions. These were those that emerged in the transition to the modern age, and with which Foucault opened *The Order of Things* (Egginton 2010; Lambert 2008; Ndalianis 2004; Maravall 1986). My gloss on this discussion is threefold.

My first gloss is to note that the condition of possibility that gives rise to the baroque, constituting what Egginton calls its very problem of thought (Egginton 2010), precipitated also by the profound shocks issuing from the discovery of the New World, was the problem space of factical finitude itself. I have, however, continually also noted that conditions of possibility find their expression in conditions of operability. My second gloss on contemporary debates concerning the baroque, 'the return of the baroque' (Lambert 2004), the 'new baroque' (Lambert 2008) and 'the ideology of neo-baroque aesthetics' (Egginton, 2010; Ndalianis 2004; Maravall 1986) is, therefore, to note, in addition, that it was the baroque that added concrete aesthetic form to the concrete political form that factical finitude took under the strategic figures of Man, and later of Life.

To move to my third observation we have, however, to move from Schmitt's omissions to Schmitt's commission. This is the additional twist that Schmitt gave to the katechon. I think it can now be seen to be at work, with full operational force, in the 'liberal way of war' (Dillon and Reid 2009) and the biopolitics of security of the twenty-first century. It is that twist which also intensifies, as it continues to characterise, the West's neo-baroque politics of security following the dissolution of the Cold War.

Arthur Bradley (2014) has been the only Anglo-American scholar, to date, to note an important nuance in Schmitt's deployment of the katechon in his account of world history. It is especially relevant here (Bradley 2014). I put this to work in a final observation about the politics of Western security during the course of the early years of the twenty-first century. What follows is indebted to Bradley's scholarship.[1]

Acceleration

Bradley reviews the introduction of the Pauline Katechon in the Apostle's Second Letter to the Thessalonians. He then reviews the reception given by Hans

Blumenberg and Jacob Taubes, in particular, to Schmitt's appeal to the katechon. In his Second Letter to the Thessalonians, Paul unpacks the sequence of events that must unfold before the return of the Messiah. First, Paul asserts that 'a lawless one' – identified by the early Church as the Antichrist – will be revealed. The Antichrist will oppose himself to God and claim to be God. The letter goes on to claim, however, that 'the mystery of lawlessness is already at work'. The only reason why the Antichrist has not appeared is because someone or something is temporarily restraining him. Paul says that the figure holding back the appearance of the Antichrist until the appointed time is the katechon:

> Do you not remember that I told you these things when I was still with you? And you know what is now restraining [*to katechon*] him, so that he may be revealed when his time comes. *For the mystery of the lawlessness is already at work, but only until the one who now restrains* [*ho katechon*] *it is removed.* And then the lawless one will be revealed, who the lord Jesus will destroy with the breath of his mouth, annihilating him by the manifestation of his coming.
>
> (Thessalonians 2:6–7, emphasis added)

This cryptic introduction of the katechon remains enigmatic in the extreme. Bradley notes that even Augustine of Hippo had trouble understanding Paul's meaning (Bradley 2014: 3). Clearly it is possible to interpret the text in a variety of ways, not least as one in which katechon and Antichrist already co-inhabit the world. The lawless one does not come. It is always already here. Its emergence is, therefore, not simply restrained by the katechon, but appears somehow to be always already enfolded within the katechon.

The restrainer (the katechon) and the restrained (the Antichrist) seem, therefore, locked in a mutually disclosive and interdependent rather than merely opposed and polemical relationship. This allows the possibility that the katechon may not simply restrain, but perhaps also accelerate the forces of salvation history. It is a thought that Schmitt seems to embrace, as he applies it to world history. I see it working throughout the 'liberal way of war' (Dillon and Reid 2009).

Thessalonians 2 makes clear that the Antichrist, or lawless one, is already at work even as, Paul also says, the katechon is already operating to retrain him. And here, Bradley invokes Esposito's interpretation as coming closest to the interdependent character of the relation between katechon and Antichrist that seems to be at work in Paul's text. The text Bradley quotes from Esposito bears repetition:

> [t]he katechon restrains evil by containing it, by keeping it at bay holding it within itself. It confronts evil, but from within, by hosting it and welcoming it, to the point of building its own necessity to the presence of evil. It limits evil, defers it, but does not eradicate it because, if it did, it would also eliminate itself. We could go so far as to say that the katechon – its constitutive juridical principle – opposes the absence of law by taking it up inside itself, and thereby, in some way, giving it form, rule, and norm.

> The katechon antinomically assigns a *nomos* to anomie, thus restraining its catastrophic unfolding.
>
> (Esposito 2011: 63)

Consider, before we move on to Schmitt's thought of the katechon as accelerator of world history, how Esposito's account of the katechon illuminates the now widespread vernacular of biopolitics of security in the twenty-first century: complex adaptive emergence, chaos, radical contingency, and resilience. These are the mechanisms by which its security politics seek to deal with the factical finitudinal threat to their end of days, assigning a nomos to the anomic and enfolding that lawlessness within itself.

As the time between the first and second coming of the Messiah became increasingly protracted, so the figure of the katechon assumed wider political significance. It served as one way of addressing the aporia of how Christians should understand and relate to secular power – specifically, in its early days, the Roman Empire – as they awaited the second coming of Christ. Tertullian, one of the early Church Fathers, argued, for example, that 'Christians should pray for the safety of the Emperor because he was the Katechon that holds back the rising tide of anarchy which would sweep Christians and pagans alike away' (Bradley 2014: 4).

As 'the time that remains' between the first and second coming began to stretch out indefinitely, however, so the katechon was employed to describe the function of state power more generally. Bradley notes that 'we find katechontic claims being made by the Byzantine, Carolingian and Holy Roman Empires' (Bradley 2014: 4). What was therefore introduced as a kind of temporary stop-gap 'in a larger eschatological story', and as the second coming was imminently awaited, begins to turn into something like 'a permanent presence in an interregnum that now stretches towards infinity' (Bradley 2014: 4).

Hermeneutical exegesis seeks to get to the truth of it, but who knows the mind of the Apostle, and who the mind of God, where the sequence of the end of days, especially, is concerned. There is no such truth to be had. But the katechon was part of a veridical apparatus that was also to become a political apparatus – a rule of truth and truth of rule. This apparatus became deeply implicated, of course, in the formation of the religio-political consciousness of the West, its institutions and practices. But, as the divine point of reference that once gave rise to it retreated into the mystery of its absolute power, so also did the katechon become detached from its original semantic field of religious reference. In the process it became a wider political apologia for secular state power.

Controversially, the Jewish thinker Jacob Taubes, for example, charged that Christian political theology became nothing less than a form of 'eschatological Caesarism' (Taubes 2004: 103). According to Bradley, it is not too much of an exaggeration to say that, for Taubes (2004), the katechon makes a Christian apology for state power of the sort exemplified in Schmitt's political theory and commitment to National Socialism (Bradley 2014: 4). But Bradley observes how the story of the

Schmittean katechon is more nuanced than Taubes allows. In Schmitt's hands, the katechon's restraining role, now transferred from salvation history to world history, begins to be powerfully complemented by the idea that it might also act as an accelerator, rather than mere restrainer, of that history.

Recalling that Schmitt's own reflections on the katechon came in short-lived bursts of writing, between the work of the 1920s and 1930s and the later work of the 1950s and 1960s, Bradley notes how Schmitt embraces different evaluations of the katechon. First adopting it as a pejorative way to signify the end of days for the British and American empires, 'it increasingly becomes a more positive conceptual descriptor in Schmitt's work' (Bradley 2014: 6). During the course of Schmitt's writings, the restrainer begins to emerge as: 'less a defensive or reactive figure than a dynamic actor in the becoming of world history: it does not merely restrain lawlessness but seeks to incorporate or internalize a kind of *anomia* into itself as its own sovereign principle' (Bradley 2014: 6).

Recounting Schmitt's account of political world history, most notably its critique of liberalism and allied attack on British and American imperialism, in promotion of the interest of Germany's Third Reich, Bradley also focuses our attention on Schmitt's differentiation of states or empires, and the character of imperial decline. Empires come and go in world history. In the process, empires seek katechontically to restrain their own decline by resisting the rise of their competitors. There are, then, for Schmitt, those that fall back into vain attempts to restrain world history, and those that accelerate it. And it is here, while reflecting in *The Nomos of the Earth* on the nature of the United States' empire, that Schmitt observes that the United States operates as 'an accelerator despite itself' (Bradley 2014: 7). In fact, it has long considered itself the accelerator by election.

My conclusion is, therefore, a simple one. Geopolitically, the modern state was classically regarded as the katechon, whose provision of security came in the form of the restrainer of chaos and lawlessness. It may also have become the container for other ambitions as well. But its katechontic task was always regarded as the condition of possibility for all other political ambitions. Katechontic restraint was once the limit condition, definitive of the state, baroquely expressed through the *arcana imperii* of modern state power. But Western geopolitics have become biopolitics in ways that classically express those complex and ornate interchanges between categories, as the sign shuttles back and forth within and between them, creating the complex textures of neo-baroque intertextuality between geopolitics and biopolitics. In its geopolitically armed pursuit of the ambition to govern the interface between life and death biopolitically as well, therefore, these politics of security give concrete political form to finitude through becoming creator and destroyer of worlds.

Notes

1 Since concluding the conclusion I discovered that William Hooker raises the question of acceleration in Chapter 5 of his book, *Carl Schmitt's International Thought*, Cambridge: Cambridge University Press, 2009.

Bibliography

Bradley, Arthur (2014) '"Accelerator Despite Itself": Credo, Crisis, Katechon' (unpublished paper).

Dean, Mitchell (2014) 'Three Forms of Modern Democratic Political Acclamation', 8th Annual Telos Conference. New York, 15 February.

Deleuze, Gilles (1992) *The Fold: Leibniz and the Baroque* (Minneapolis: University of Minnesota Press).

Dillon, Michael and Reid, Julian (2009) *The Liberal Way of War: Killing to Make Life* (Abingdon: Routledge).

Egginton, William (2010) *The Theater of Truth: The Ideology of (Neo)Baroque Aesthetics* (Stanford, CA: Stanford University Press).

Esposito, Roberto (2011) 'The Katechon', in Esposito, *Immunitas: The Protection and Negation of Life* (Cambridge: Polity Press), pp. 52–79.

Foucault, Michel (2013) *Speech Begins after Death*, ed. Philippe Artières (Minneapolis: University of Minnesota Press).

Hobbes, Thomas (1960) *Leviathan*, ed. Michael Oakeshott (Oxford: Blackwell).

Lambert, Greg (2004) *The Return of the Baroque in Modern Culture* (London: Continuum).

Lambert, Greg (2008) *On the (New) Baroque* (Aurora, CO: Davis Group).

Maravall, José Antonio (1986) *Culture of the Baroque: Analysis of a Historical Structure*, trans. Terry Cochran (Minneapolis: University of Minnesota Press.

Ndalianis, Angela (2004) *(Neo)Baroque Aesthetics and Contemporary Entertainment* (Cambridge, MA: MIT Press).

Taubes, Jacob (2004) *The Political Theology of Paul* (Stanford, CA: Stanford University Press).

INDEX